THE BALUNGI LECTURES ON PHYSICS FOR THE USE OF SCHOOLS AND COLLEGES

Mainly Dark Matter, Black Holes, Quantum Mechanics, General Relativity and Quantum Gravity

BALUNGI FRANCIS

Author of Quantum Gravity in a Nutshell1

https://www.balungifrancisbooks.com

bfrancis@cedat.mak.ac.ug, balungif@gmail.com

Learn from the Fathers of Physics

https://www.zhizheloo.com

Copyright ©2021 Balungi Francis

Copyright ©2021Bill Stone Services

Balungi Francis asserts the moral right to be identified as the author of this work.

All rights reserved. Apart from any fair dealing for the purposes of research or private study or critism or review, no part of this publication may be reproduced, distributed, or transmitted in any form or by any means, including photocopying, recording, or other electronic or mechanical methods, or by any information storage and retrieval system without the prior written permission of the publisher.

TABLE OF CONTENTS

DEDICATION 1

PREFACE   2

1. An Exceptionally Simple Theory of Gravity 3

2. Space-time Singularity Resolution in Quantum Gravity   16

3. The Minimum mass limit of a gravitationally Collapsed star   26

5. Reinventing Gravity 44

6. The Simple Link between Quantum Mechanics and Gravity   78

7. Resolution of the Proton Radius Puzzle   87

8. On the Irreducible Anomaly in Einstein Deflection Angle   93

9. A New Approach to Quantum Theory   105

10. Vacuum Pressure Gravity as a Possible Alternative to Newton's law of Gravity 121

11. The Theory of Everything   125

12. What is Semi-Classical Gravity?   142

13. The Extra Dimension Problem   158

14. How can the Laws of Physics be derived from one Underlying Principle?   173

15. Is Gravity and the Laws of Physics Emergent?   180

16. The Galaxy Rotation Problem   198

17. Particle Creation by Black Holes: Is it Hawking's Approach or My Approach?   214

18. The Spectrum of the Atomic Universe   232

19. On the Generalization of Loop Quantum Gravity 241

20. An Exceptionally Simple Classical Unified Field Theory   247

21. The earliest period of time in the history of the Universe   252

22. Another form of general relativity and its new predictions 259

23. Balungi's 2010 Research in Quantum Gravity   276

Bibliography   292

Acknowledgments   313

About the Author   314

Glossary   316

DEDICATION

To my lovely mother Mukajjuna Annet, here is your Nobel Prize

To my wife W. Ritah for her constant feedback throughout and many long hours of editing,

To my sons Odhran and Leander,

To Carlo Rovelli, Lee Smolin, Neil deGrasse Tyson and Sabine Hossenfielder, I say thank you for your astonishing suggestions.

PREFACE

This is a National bestselling physics textbook based on some lectures by Balungi Francis, a Quantum Gravity researcher who has sometimes been called —The Next Einstein‖. The lectures were presented before graduate students at the College of Engineering, Design, Art and Technology (Makerere), during 2012-2014, Ranging from the most basic principles of Newtonian physics through such formidable theories as general relativity and quantum mechanics, explaining dark matter and developing a quantum theory of gravity. Balungi's lectures stand as a monument of clear exposition and deep insight. Timeless and collectible, the lectures are essential reading, not just for students of physics but for anyone seeking an introduction to modern physics. Not since the Feynman lectures on physics has physics been so vividly, intelligently and entertainingly revealed.

Balungi Francis

March, 17, 2021

1. An Exceptionally Simple Theory of Gravity

Re-thinking General Relativity

Seventeen years ago I made what at first seemed like a small discovery: a vacuum pressure gravitational theory of mine deduced something I did not expect. But the more I investigated, the more I realized that what I had seen was the beginning of a crack in the very foundations of existing physics, and a first clue towards a whole new kind of physics.

This book is the culmination of nearly seventeen years of work that I have done to develop that new kind of physics. I had never expected it would take anything like as long, but I have discovered vastly more than I ever thought possible, and in fact what I have done now touches almost every existing problem in physics.

In the early years, I published some papers in the major scientific research journals which were well received but because they had become scattered, I resolved just to keep working quietly until I had finished, and was ready to present everything in a single coherent way. Thirteen years later this book is the result. And with it my hope is to share what I have done with a wide range of scientists and non-scientists as possible.

And now that I have finished building the intellectual structure that I describe in this book, it is my hope that those who read these words can share in the excitement I have had in making the discoveries that were involved.

In this section we propose a new theory of gravity. This theory states that, gravity is a result of the vacuum energy density fluctuations. This theory has been proven beyond reasonable doubt and agrees with existing observations.

1. Propositions

1) Vacuum space is not empty. The vacuum is fluctuating and generating particle pairs that appear and disappear incredibly quickly.

2) There might be a medium with physical properties filling 'empty' space, an aether, enabling the observed physical processes. Therefore aether is a constant energy density filling space homogeneously

3) Aether=Dark energy=zero-point energy=vacuum energy density

2. Historical Conjectures and proposals

1)Isaac Newton in 1704: "Doth not this aethereal medium in passing out of water, glass, crystal, and other compact and dense bodies in empty spaces, grow denser and denser by degrees, **and by that means refract the rays of light not in a point, but by bending them gradually in curve lines?** ...Is not this medium much rarer within the dense bodies of the Sun, stars, planets and comets, than in the empty celestial space between them? And in passing from them to great distances, doth it not grow denser and denser perpetually, and thereby cause the gravity of those great bodies towards one another, and of their parts towards the bodies; every body endeavouring to go from the denser parts of the medium towards the rarer?"

2)Albert Einstein in 1894 or 1895: "The velocity of a wave is proportional to the square root of the elastic forces which cause [its] propagation, and inversely proportional to the mass of the aether moved by these forces."

3)Albert Einstein in 1920: "There is a weighty argument to be adduced in favour of the aether hypothesis. **To deny the aether is ultimately to assume that empty space has no physical qualities whatever.** The fundamental facts of mechanics do not harmonize with this view... according to the general theory of relativity space is endowed with physical qualities; in this sense, therefore, there exists an aether. **According to the general theory of relativity space without aether is unthinkable; for in such space there not only would be no propagation of light, but also no possibility of existence for standards of space and**

time (measuring-rods and clocks), nor therefore any space-time intervals in the physical sense. But this aether may not be thought of as endowed with the quality characteristic of ponderable media, as consisting of parts which may be tracked through time. The idea of motion may not be applied to it.it."

3)Robert B. Laughlin in 2005: "It is ironic that Einstein's most creative work, the general theory of relativity, should boil down to conceptualizing space as a medium when his original premise [in special relativity] was that no such medium existed [..] The word 'ether' has extremely negative connotations in theoretical physics because of its past association with opposition to relativity. This is unfortunate because, stripped of these connotations, it rather nicely captures the way most physicists actually think about the vacuum. . . . Relativity actually says nothing about the existence or nonexistence of matter pervading the universe, only that any such matter must have relativistic symmetry. [..] It turns out that such matter exists. About the time relativity was becoming accepted, studies of radioactivity began showing that the empty vacuum of space had spectroscopic structure similar to that of ordinary quantum solids and fluids. Subsequent studies with large particle accelerators have now led us to understand that space is more like a piece of window glass than ideal Newtonian emptiness. It is filled with 'stuff' that is normally transparent but can be made visible by hitting it sufficiently hard to knock out a part. The modern concept of the vacuum of space, confirmed every day by experiment, is a relativistic ether. But we do not call it this because it is taboo."

3. Basic Theory

Gravity is a result of vacuum energy density, a constant energy density filling space homogeneously.

The attraction between two bodies (e.g Sun-Earth attraction) is proportional to:

$$F = \sqrt{4\pi\alpha\hbar c\rho_v} \quad (1)$$

Where α is the coupling constant which determines the strength of a force in an interaction, $\hbar$ is the reduced Planck constant c is the velocity of light particles, and ρ_v is the density of the medium (call it aether, dark matter etc).

The above given equation means that: the vacuum is not empty but consists of particles popping in and out of existence. This vacuum therefore has an energy density ρ_v which leads to the reaction force that causes the motion of matter in the vacuum- non empty space. It is this force we perceive as gravity:

4. Curvature of light-rays. Perihelion-motion of the paths of the Planets

Let us find out the curvature which a light-ray suffers when it goes by a mass M at a distance R from it. If we use the notion of vacuum energy density fluctuations, then the total bending B of light-rays (reckoned positive when it is concave to the origin) is given as a sufficient approximation of the ratio of the Fluctuation energy density

ρ_M of a body of mass M (e.g Sun) to the vacuum energy density Fluctuation ρ_v filling all space by

$$B = \frac{\rho_M}{\rho_v}$$

The vacuum pressure P in a circular body of mass M and area $A = \pi R^2$ where R is the radius of M, is

$$P = \rho_M = \frac{F}{\pi R^2}$$

On putting in equation (1) and arranging we have:

$$\frac{\rho_M}{\rho_v} = \frac{4\alpha\hbar c}{\pi \rho_M R^4}$$

The coupling constant that determines the strength of the gravitational force is:

$$\alpha = \frac{GM^2}{\hbar c}$$

Also the total energy density of the body of mass M with volume $V \sim \pi R^3$ (consider Einstein mass-energy conversion)

$$\rho_M = \frac{Mc^2}{\pi R^3}$$

Finally on substitution, the calculation gives:

$$B = \frac{\rho_M}{\rho_v} = \frac{4GM}{Rc^2}$$

A ray of light just grazing the sun would suffer a bending of $1.75 arcsec$. If we calculate the gravitation-field to a greater order of approximation and with it the corresponding path of a material particle of a relatively small (infinitesimal) mass we set a deviation of the following kind from the KEPLER-NEWTONian Laws of Planetary motion.

5. Newton's theory as a first approximation

Consider two bodies M and m in a vacuum of space. Draw a line connecting M to m with a distance R apart. Then draw a sphere around M such that m rests on this sphere with M at the center of the sphere. Let R be the radius of the sphere, where the surface area of the sphere is $A = 4\pi R^2$. Then the vacuum pressure/energy density fluctuations within this sphere is calculated to be,

$$P = \rho_v = \frac{F}{\pi R^2}$$

On substitution into equation (1) we have

$$F = \frac{\alpha \hbar c}{R^2}$$

The coupling constant that determines the strength of the gravitational force is:

$$\alpha = \frac{GMm}{\hbar c}$$

Where G denotes the gravitational constant, on substitution we get the usual law of universal gravitation as,

$$F = \frac{GMm}{R^2}$$

6. Calculation of the Energy stored in the Electromagnetic field

The electric force on a particle of charge e in the electric field E is

$$F = Ee$$

The coupling constant that determines the strength of the electromagnetic force is:

$$\alpha = \frac{e^2}{4\pi\varepsilon_0 \hbar c}$$

Putting all of these into consideration (in equation1) we have:

$$\rho = \varepsilon_0 E^2$$

This is the usual formula for the energy density stored in the electromagnetic field. Therefore eqn1 becomes a general rule for all fields whether electromagnetic, gravitational etc. It is the basis for our study of the origin of the fundamental forces of nature.

7. Calculation of the cosmological constant Zero point vacuum energy density

The discovery that most of the mass of the universe is accounted for by an unknown form of matter called dark matter and that the expansion of the universe is accelerating, which has been attributed to a form of energy called dark energy has stimulated renewed interest in the question of quintessence. It has been suggested by some authors [1] [2] [3] that a form of dark energy called quintessence could be a fifth fundamental force.

This sounds strange because in physics there are four observed fundamental forces or interactions that form the basis of all known interactions in nature: gravitational, electromagnetic, strong nuclear, and weak nuclear forces. Although these forces work well in their domain of applicability, they seem to fail in others, for example; they fail to explain what dark matter and dark energy really is. This therefore raises questions as to whether there is a classical formula of the force of the dark energy like that of the gravitational force? Dark energy's status as a hypothetical force with unknown properties makes it a very active target of research.

In cosmology, dark energy has been explained by the **cosmological constant** -the energy density of space, or vacuum energy. However, there is one problem, the cosmological constant problem, which has been termed as One of "the worst theoretical prediction in the history of physics" and "the largest discrepancy between theory and experiment in all of science", it is the disagreement between the observed values of vacuum energy density (the small value of the cosmological constant) and theoretical large value of zero-point energy suggested by quantum field theory.

Attempts to find a way to measure or to calculate the value of the vacuum energy density and the cosmological constant have all either failed or produced results incompatible with observations or other confirmed theoretical results. Some of those results are theoretically implausible because of certain unrealistic assumptions on which the calculation model is based. And some theoretical results are in conflict with observations, the conflict itself being caused by certain questionable hypotheses on which the theory is based. And the best experimental (Casmir effect) evidence is based on the measurement of the difference of energy density within and outside of the measuring apparatus, thus preventing in principle any numerical assessment of the actual energy density. In this paper I have proposed a fifth fundamental force to explain various anomalous observations that do not fit existing theories. I believe that this fifth force can led to the observed cosmological constant.

Consider a galaxy that is held together by unknown form of matter "dark matter", and accelerating away from us due to the accelerating expansion of the universe "dark energy". This is a difficult problem that cannot be explained by existing theories. But it can be summed up and explained by this simple definition of the fifth fundamental force of nature as;

From Newton's second law of motion, the fifth force is a product of the mass M_{pl} of a Planck particle (the hypothetical atom of space in loop quantum gravity) and the acceleration a_o (low acceleration in MOND). This force is also equivalent to the Mass of a Planck particle and the Hubble constant H_o by,

$$F = M_{pl}a_o \sim M_{pl}cH_o\sqrt{\frac{3}{2}\alpha\Omega}$$

From equation (1), the vacuum energy density is given by,

$$\rho = \frac{F^2}{4\pi\alpha\hbar c}$$

The coupling constant that determines the strength of the gravitational force is proportional to the square of the mass of a dark matter particle as:

$$\alpha = \frac{GM^2}{\hbar c}$$

Which gives the critical energy density value;

$$\rho = \frac{a_o^2}{4\pi G}\left(\frac{M_{pl}}{M}\right)^2 = \frac{3H_o^2 c^2}{8\pi G}\Omega = 5.336 \times 10^{-10}\,\frac{J}{m^3}$$

Where the verified parameters are calculated to be;

1) Low acceleration in Modified newtonian dynamics

$$a_o = \frac{a_{Milogram}}{\sqrt{2}} = 8.485281 \times 10^{-11} ms^{-2}$$

2) Mass of a dark matter particle

$$M = 2.76597 \times 10^{-9} kg$$

In conclusion we have just discovered that, there exists a fifth fundamental force of nature. The particle attributed to this force is the Planck mass particle Y20 of a Planck length size which is also the particle of dark matter according to the theory given above. The fifth force is therefore responsible for the accelerated expansion of the universe attributed to dark energy and we are yet to provide an experimental verification of this force in the coming papers.

2.Space-time Singularity Resolution in Quantum Gravity

It has been known for some time that a star more than three times the size of our Sun collapses in this way, the gravitational forces of the entire mass of a star overcomes the electromagnetic forces of individual atoms and so collapse inwards. If a star is massive enough it will continue to collapse creating a Black hole, where the whopping of space time is so great that nothing can escape not even light, it gets smaller and smaller. The star in fact gets denser as atoms even subatomic particles literally get crashed into smaller and smaller space, and its ending point is of course a space time singularity.

In summary, a Black hole is that object created when a dying star collapses to a singular point, concealed by an event horizon, it is so dense and has strong gravity that nothing, including light, can escape it. Black holes are predicted by general relativity, and though they cannot be "seen," several have been inferred from astronomical observations of binary stars and massive collapsed stars at the centers of galaxies.

Black holes formed by gravitational collapse require great energy density but there exists a new breed of Black holes that where formed in the early universe after the big bang, where the energy density was much greater allowing the formation of Primordial Black holes with masses ranging from, $10^8, 10^{12} - 10^{17} kg$. Therefore the formation of primordial, min or quantum black holes was due to density perturbations forming in it a gravitational collapse in the early universe.

A Black hole might not actually be a physical object in space but rather a mathematical singularity, a prediction of Einstein's General Relativity theory, a place where the solutions of Einstein differential equations break down. A space-time singularity therefore is a position in space where quantities used to determine the gravitational field become infinite; such quantities include the curvature of space-time and the density of matter. Singularities are places where both the curvature and the energy-density of matter become infinitely large such that light cannot escape them. This happens for example inside black holes and at the beginning of the early universe.

Singularities in any physical theory indicate that either something is wrong or we need to reformulate the theory itself. Singularities are like dividing something by zero.

The problems that plague the General relativity theory arise from trying to deal with a point in space or a universe that is zero in size (infinite densities). However, quantum mechanics suggests that there may be no such thing in nature as a point in space-time, implying that space-time is always smeared out, occupying some minimum region. The minimum smeared-out volume of space-time is a profound property in any quantized theory of gravity and such an outcome lies in a widespread expectation that singularities will be resolved in a quantum theory of gravity. This implies that the study of singularities acts as a testing ground for quantum gravity.

Loop quantum gravity (LQG) suggests that singularities may not exist. LQG states that due to quantum gravity effects, there must be a minimum distance beyond which the force of gravity no longer continues to increase as the distance between the masses become shorter or alternatively that interpenetrating particle waves mask gravitational effects that would be felt at a distance. It must also be true that under the assumption of a corrected dynamical equation of LQ cosmology and brane world model, for the gravitational collapse of a perfect fluid sphere in the commoving frame, the sphere does not collapse to a singularity but instead pulsates between a maximum and minimum size, avoiding the singularity.

Additionally, the information loss paradox is also a hot topic of theoretical modeling right now because it suggests that either our theory of quantum physics or our model of black holes is flawed or at least incomplete. and perhaps most importantly, it is also recognized with some prescience that resolving the information paradox will hold the key to a holistic description of quantum gravity, and therefore be a major advance towards a unified field theory of physics.

Singularities are a sign that the theory breaks down and has to be replaced by a more fundamental theory. And we think the same has to be the case in General Relativity, where the more fundamental theory to replace it is quantum gravity.

Whether in gravitational collapse or the early universe, we now know that the formation of Black holes or space time singularities requires great and much greater energy density. This we know because while the left hand side of Einstein field equations representsnts the metric of space-time curvature, the right hand side represents the matter-energy content of the classical matter fields of pressure and energy density. This therefore means that quantum mechanics which plays an important role in the behavior of the matter fields has no place in the Einstein field equations and this is what brings on the singularities that plague the general relativity theory.

$$G_{\mu\nu} + \Lambda g_{\mu\nu} = \frac{8\pi G}{c^4} T_{\mu\nu}$$

Because of this, one therefore has a problem of defining a consistent scheme in which the space time metric is treated classically but is coupled to the matter fields which are treated quantum mechanically.

The approximation I shall use on my journey to quantum gravity (Quantum Black holes) is that the matter fields, such as scalar, electro-magnetic, or neutrino fields, obey the usual wave equations with the left hand side replaced by a classical space time second order curvature

$$\Lambda = \frac{1}{R^2}$$

Where R is the radius of curvature, while the right hand stress-energy tensor is replaced by the quantum mechanical energy density (see first chapter)

$$\rho = \frac{F^2}{8\pi\alpha\hbar c} \qquad (1)$$

Where F is the force involved in an interaction α is the coupling constant that determines the strength of the force, and ℏ is the reduced Planck constant. The equation represents the coupling constant (α) as a function of the energy density (ρ) for any force (F) exerted in an interaction. The application of this equation is the Franzl Aus Tirol curve on Wikipedia's "Coupling constant". Another application is the derivation of energy stored in the electromagnetic field. Therefore the general theory of quantum mechanics in curved space –time will be given by this simple equation,

$$\Lambda = \frac{8\pi G}{c^4}\rho$$

Where,

$$\Lambda = \frac{GF^2}{\alpha\hbar c^5} = \frac{F^2}{\alpha E_{pl}^2} \qquad (2)$$

Where, $E_{pl} = M_{pl}c^2$ is the Planck energy and M_{pl} is the Planck mass

From the above given equation we see that high space curvature will always be achieved when the square of the force involved increases. According to the theory given, this will only occur at the Planck energy level where space is discrete or granular in nature (its building blocks being exactly the Planck mass, simply put, the atoms of space). There is no change in energy because the only energy involved in the process is the constant Planck energy of the Planck mass.

As we said earlier, that the formation of a black hole due to the process of gravitational collapse occurs in the presence of great energy density and also that the formation of primordial black holes in the early universe occurs in the presence of a much greater energy density, our theory suggests that this energy density is high because of the strong gravitational force involved in the process. According to general relativity, this force is a constant and is given by

$$F = \frac{c^4}{G}$$

Therefore from equation (2), when this force is present the curvature of space scales as the inverse of the square of the Planck length,

$$\Lambda = \frac{c^3}{\alpha \hbar G} = \frac{1}{\alpha l_p{}^2} \qquad (3)$$

Where $l_p = \sqrt{\frac{\hbar G}{c^3}}$ is the Planck length.

This implies that, in the theory of quantum mechanics in curved space-time for the gravitational collapse of a star, the star does not collapse to a singularity but instead to a Planck sized star of Planck length close to $10^{-35} m$ and this will happen only when $\alpha = 1$. Finally, in the theory of quantum mechanics in curved space-time, we consider the possibility that the energy of a collapsing star and any additional energy falling into the hole could condense into a highly compressed core with density of the order of the Planck density. Since the energy density or pressure is expressed as in equation (1),

$$\rho = \frac{F^2}{8\pi \alpha \hbar c}$$

Therefore nature appears to enter the quantum gravity regime when the energy density of matter reaches the Planck scale. The point is that this may happen well before relevant lengths become planckian. For instance, a collapsing spatially compact universe bounces back into an expanding one. The bounce is due to a quantum-gravitational repulsion which originates from the modified Heisenberg uncertainty, and is akin to the force that keeps an electron from falling into the nucleus. And from the uncertainity principle, this repulsion force is given by,

$$F = \frac{c^4}{G}$$

Therefore the bounce does not happen when the universe is of planckian size, as before; it happens when the matter energy density reaches the Planck density in this way,

$$\rho = \frac{c^7}{8\pi\alpha\hbar G^2} \quad (4)$$

At this energy density, a Planck star is formed. The key feature of this theoretical object is that this repulsion arises from the energy density, not the Planck length, and starts taking effect far earlier than might be expected. This repulsive 'force' is strong enough to stop the collapse of the star well before a singularity is formed, and indeed, well before the Planck scale for distance. Since a Planck star is calculated to be considerably larger than the Planck scale for distance, this means there is adequate room for all the information captured inside of a black hole to be encoded in the star, thus avoiding information loss.

The analogy between quantum gravitational effects on Cosmological and black-hole singularities has been exploited to study if and how quantum gravity could also resolve the r = 0 singularity at the center of a collapsed star, and there are good indications that it does. For example, if we extend (3) to n extra dimensions we have,

$$R = \alpha^{n/2} l_p$$

Where α in this case is the size of the extra dimensions and α^n is the flux in the extra dimesions. Let the size of the extra dimension be given as the gravitational coupling constant, $\alpha = \frac{GM^2}{\hbar c} = \left(\frac{M}{M_{pl}}\right)^2$, then the size of a star will be given by,

$$r = \left(\frac{M}{M_{pl}}\right)^n l_p \qquad (5)$$

Where M is the mass of the star and n is positive. For instance, if n = 1/3, a stellar-mass black hole would collapse to a Planck star with a size of the order of 10^{-10} centimeters. This is very small compared to the original star in fact, smaller than the atomic scale but it is still more than 30 orders of magnitude larger than the Planck length. This is the scale on which we are focusing here. The main hypothesis here is that a star so compressed would not satisfy the classical Einstein equations anymore, even if huge compared to the Planck scale because its energy density is already Planckian.

3. The Minimum mass limit of a gravitationally Collapsed star

The smallest black hole would be one where the Schwarzschild radius equals the radius of a mass with a reduced Compton wavelength which is the smallest size to which a given mass can be localized. For a small mass M, the Compton wavelength exceeds half the Schwarzschild radius, and no black hole description exists. This smallest mass for a black hole is thus approximately the Planck mass, the micro black hole.

Contrary to the above observation, torsion (see Einstein-Cartan theory) modifies the Dirac equation in the presence of the gravitational field causing fermions to be spatially extended. This spatial extension of fermions limits the minimum mass of a black hole to be on the order of 10^{16} Kg, showing that micro black holes (of Planck mass) may not exist. Another mass limit is from the data of the Fermi Gamma-ray space telescope satellite which states that, less than one percent of dark matter could be made of primordial black holes with masses up to 10^{13} Kg.

The major aim of this book is to prove theoretically the existence of a minimum mass limit of a gravitationally collapsed star and thereafter prove **Chandrasekhar** wrong (see Chandrasekhar 1983 Noble lecture concluding statement below)

"We conclude that there can be no surprises in the evolution of stars of mass less than 0.43Solarmass ($\mu = 2$). The end stage in the evolution of such stars can only be that of the white dwarfs. (Parenthetically, we may note here that the so-called 'mini' black-holes of mass 10^{12} Kg cannot naturally be formed in the present astronomical universe.)"

In what follows the above given statement may prove to wrong according to a detailed derivation given below.

From the theory of white dwarf stars, the radius limit of a white dwarf of mass M is given by the following equation,

$$R_w = \frac{(9\pi)^{2/3}}{8} \frac{\hbar^2}{m_e G (m_{pro})^{5/3} M^{1/3}} \quad (1)$$

Where m_{pro} and m_e is the proton and electron mass respectively

Just like the Compton wavelength, there must exist another radius for the consistitution of stars that differs from the radius given in (1) above. For example, in the same way the Planck mass is deduced (i.e by equating the Schwarzschild radius to the Compton wavelength) is the same way in which we are to prove the existence of the mass limit of a gravitationally collapsed star.

We start from first principles. Let it be known that the derivation of the Chandrasckhar mass limit will follow the equipartition of the gravitational potential energy of a star to its electron degeneracy pressure. In the same way, if the gravitational binding energy is given by,

$$E_g = \frac{2 M_{pl}^3 m_e c^2 (6.144\pi^3)}{M m_{pro}^2 \; ?^2} \quad (2)$$

Where M_{pl} is the Planck mass and μ is is the average molecular weight per electron

And the electron degeneracy energy pressure of the star is given by,

$$E_d = m_e c^2$$

When $E_g = E_d$ then we obtain the mass limit of the white dwarf star as,

$$M = \frac{12.288\pi^3 M_{pl}^3}{\mu_e^2 \; m_p^2} = 1.4 M_{sun}$$

If then this is true, then the formula (2) for the gravitational binding energy of a star is true. This therefore implies that the following assumption will also be true.

When the binding gravitational energy of a star is equal to the Newtonian gravitational potential energy $\frac{GM^2}{R}$ we obtain the radius which is the smallest size to which a given mass of a star can be localized as,

$$\frac{GM^2}{R} = \frac{2M_{pl}{}^3 m_e c^2 (6.144\pi^3)}{M m_{pro}{}^2 \, \mu^2}$$

$$R = \frac{G m_{pro}{}^2}{2 m_e c^2} \left(M/M_{pl}\right)^3 \frac{\mu^2}{6.144\pi^3} \quad (3)$$

This can be rewritten in the form,

$$R = R_k \left(M/M_{pl}\right)^3$$

Where $R_k = 2.384 \times 10^{-53}$ m which is smaller than the Planck length of 1.62×10^{-35} m

Therefore equating Equation (1) to Equation (3) we deduce the mass limit of a gravitationally collapsed star as,

$$M = \left(\frac{293.534 \pi^{11} M_{pl}{}^{21}}{\mu^6 M_{pro}{}^{11}}\right)^{1/10}$$

$$= 9.54 \times 10^{13} \, Kg$$

The value is in excellent agreement with other theoretical and experimental observations

The radius of this black hole from Equation (3) is thus 2.527×10^{14} m larger than the radius of the sun of 7×10^{8} m.

In conclusion therefore the end stage in the evolution of a star can only be that of the black hole with a mass 9.54×10^{13} Kg and size of 2.53×10^{14} m in contrast with the Chandrasekhar observations.

Note that the radius given by Equation (3), $R = R_k \left(M/M_{pl} \right)^3$ above is similar to the Equation for the size of the Planck star that was given by Rovelli and Vidotto, $r = l_p \left(\frac{M}{M_{pl}} \right)^n$ where l_p is the Planck length and n is the positive number. This is a clear indication that singularities in black holes can be resolved.

4. The Volume Entropy Law of a Black Hole

The development of general relativity followed a publication of acceleration under special relativity in 1907 by Albert Einstein. In his article, he argued that any mass will "Distort" the region of space around it so that all freely moving objects will follow the same curved paths curving toward the mass producing the distortions. Then in 1916, Schwarzschild found a solution to the Einstein field equations, laying the groundwork for the description of gravitational collapse and eventually black holes.

A black hole is created when a dying star collapses to a singular point, concealed by an "event horizon;" the black hole is so dense and has such strong gravity that nothing, including light, can escape it; black holes are predicted by general relativity, and though they cannot be "seen," several have been inferred from astronomical observations of binary stars and massive collapsed stars at the centers of galaxies.

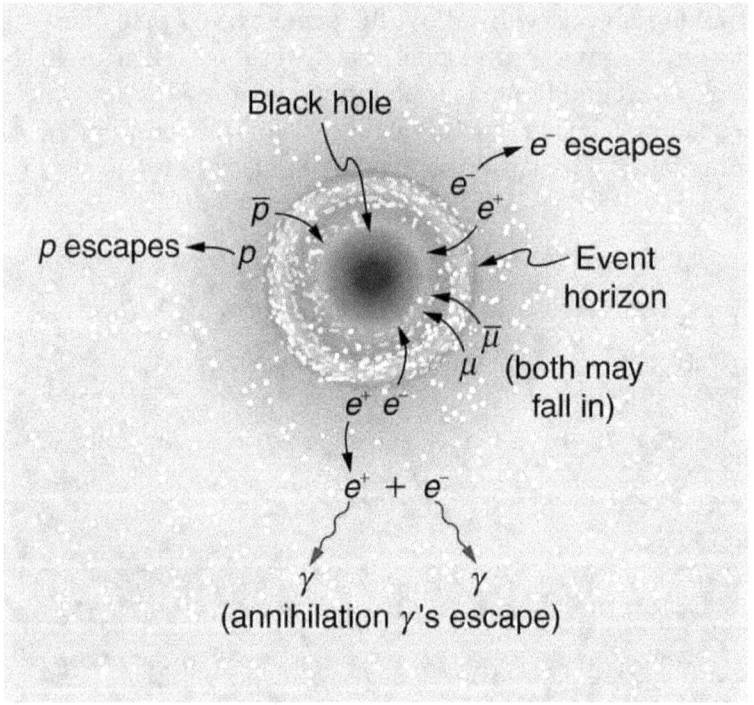

These objects have puzzled the minds of great thinkers for many years. History puts it that, they were first predicated by John Michell and Pierre-Simon Laplace in the 18th century but David Finkelstein was the first person to publish a promising explanation of them in 1958 based on Karl Schwarz child's formulations of a solution to general relativity that characterized black holes in 1916.

In 1971, Hawking developed what is known as the second law of black hole mechanics in which the total area of the event horizons of any collection of classical black holes can never decrease, even if they collide and merge. This was similar to the second law of thermodynamics which states that, the entropy of a system can never decrease.

Then in 1972 Bekenstein proposed an analogy between black hole physics and thermodynamics in which he derived a relation between the entropy of black hole entropy and the area of its event horizon.

$$S = \frac{Akc^3}{4G\hbar}$$

In 1974, Hawking predicted an entirely astonishing phenomenon about black holes, in which he ascertained with accuracy that black holes do radiate or emit particles in a perfect black body spectrum.

$$T = \frac{\hbar c^3}{8\pi G M k}$$

The Bekenstein-Hawking area entropy law raises a number of questions. Why does the entropy of a Black hole scale with its area and not with its volume? For systems that we have studied, the entropy is proportional to the volume of the system. If entropy is proportional to area, so what do we make of all those thermodynamic relations that include volume, like Boyle's law or descriptions for a gas in a box? In otherwords how do we associate volume to the entropy of a Black hole?

Area Entropy Law

A Black hole is defined as a mathematical spacetime singularity that is; a position in space where quantities used to determine the gravitational field become infinite; such quantities include the curvature of spacetime and the density of matter. That is, for high or infinite densities where matter is enclosed in a very small volume of space General relativity breaks down.

Quantum mechanics suggests that there may be no such thing in nature as a point in space-time, implying that space-time is always smeared out, occupying some minimum region. The minimum smeared-out volume of space-time is a profound property in any quantized theory of gravity and such an outcome lies in a widespread expectation that singularities will be resolved in a quantum theory of gravity. Therefore associating area to the entropy of a black hole means that the black hole has no volume which may not be true in a theory of quantum mechanics in curved space-time.

If density is the amount of energy contained within a given volume of space, then a Black hole must have a density and its volume will be determined by the amount of space enclosed by the surface area of its event horizon. Then from our energy density equation (see previous series) we can derive the Area entropy law of a Black hole. Let the intensity I of the radiation emitted by a black hole when the amount of energy is added be

$$I = \frac{F^2}{8\pi\hbar}$$

Intensity is power per unit surface area of a Black hole A and power is the amount of energy E added to a black hole in time t. This implies that the energy or heat added to a black hole is

$$E = \frac{AtF^2}{2\pi\hbar}$$

Where ℏ is the reduced Planck constant

From Einstein General relativity, the gravitational force felt by a body near the black hole is very strong and is given by,

$$F = \frac{c^4}{4G}$$

From the time-energy Uncertainty Principle, the time taken by the black hole of mass M to dissipate is,

$$t = \frac{\hbar}{Mc^2}$$

Then on substitution into the energy formula we have

$$W = \frac{Ac^6}{32\pi G^2 M}$$

This is the Frodden-Ghosh- Perez Energy.

Hawking knew that if the horizon area were an actual entropy, black holes would have to radiate. When heat is added to a thermal system, the change in entropy is the increase in thermal energy divided by temperature:

$$S = \frac{E}{T}$$

From Hawking original calculation, the temperature of a black hole is given by,

$$T = \frac{\hbar c^3}{8\pi GMk}$$

Therefore the Entropy of a Black hole will be given by,

$$S = \frac{\left(\frac{Ac^6}{32\pi G^2 M}\right)}{\left(\frac{\hbar c^3}{8\pi GMk}\right)} = \frac{Akc^3}{4G\hbar}$$

This implies that the heat added into the black hole goes into increasing its area and an increase in area will automatically lead to an increase in entropy as explained by the above given area entropy law.

Black hole Volume Entropy law

If density is the amount of energy contained within a given volume of space, then a Black hole must have a density and its volume will be determined by the amount of space enclosed by the surface area of its event horizon. Let the energy density of a black hole be given as the work done on the system by exterior agents (E) per unit volume (V),

$$\rho = \frac{E}{V} = \frac{F^2}{2\pi\hbar c}$$

(Note: Instead of the Intensity formula we have used the energy density formula)

As before, the strong gravitational force of a black hole is

$$F = \frac{c^4}{4G}$$

The energy added to the black hole is related to the Volume of the black hole by;

$$E = \frac{Vc^7}{32\pi\hbar G^2}$$

This energy is very different from the Frodden-Ghosh-Perez Energy. This means that the energy added goes into increasing the volume of the Black hole not its area.

Hawking knew that if the horizon area were an actual entropy, black holes would have to radiate. When heat is added to a thermal system, the change in entropy is the increase in thermal energy divided by temperature:

$$S = \frac{E}{T}$$

From Hawking original calculation, the temperature of a black hole is given by,

$$T = \frac{\hbar c^3}{8\pi GMk}$$

Therefore the Entropy of a Black hole will be given by,

$$S = \frac{\left(\frac{Vc^7}{32\pi \hbar G^2}\right)}{\left(\frac{\hbar c^3}{8\pi GMk}\right)} = \frac{VkMc^4}{4\hbar^2 G}$$

Therefore the entropy of a black hole is related to the Volume and mass of a black hole by the above given formula. Therefore an increase in both mass-energy Mc^2 and volume of the black hole leads to an increase in entropy as

$$S = V(Mc^2)\frac{kc^2}{4\hbar^2 G}$$

However in a limit where the Volume of the black hole is related to its area A, that is for Black holes the size of Debrogile wavelength $\lambda = \frac{\hbar}{Mc}$

$$V = \frac{\hbar A}{Mc}$$

We recover the Bekenstein Hawking area entropy law.

Finally when the black hole has a mass equal to the Planck mass particle

$$M = \sqrt{\frac{\hbar c}{G}}$$

We get the entropy of the black hole to be

$$S = \frac{Vk}{4}\left(\frac{c^3}{G\hbar}\right)^{3/2}$$

5. Reinventing Gravity

Just as Einstein in his day was constructing an alternative gravity theory, an alternative to Newton's law of gravity,which had prevailed for more than 200 years, so I have been searching for a larger theory, a modification of general relativity that would fit the data without the need to posit dark matter, and would contain Einstein's theory just as Einstein contains Newton. Unlike the alternative gravity theories known today, my mature Modified Gravity theory, contains no physical instabilities. It is as robust a gravity theory as general relativity, and fits all the current astrophysical and cosmological data without dark matter.

Modified Newtonian Dynamics (MOND) by Milogram predicted a modified Newtonian acceleration law that could fit the large amount of anomalous rotational velocity curve data from galaxies obtained by astronomers since the late 1970s, which showed stars rotating at the edges of galaxies at twice the rate predicted by Newton and Einstein. My aim is to try to explain the data without the conventional reliance on exotic dark matter.

In Newton's view, all objects exert a force that attracts other objects. That universal law of gravitation worked pretty well for predicting the motion of planets as well as objects on Earth and it's still used, for example, when making the calculations for a rocket launch. But Newton's view of gravity didn't work for some things, like Mercury's peculiar orbit around the sun. The orbits of planets shift over time, and Mercury's orbit shifted faster than Newton predicted. In spiral galaxies, the orbiting of stars around their centers seems to strongly disobey Newton's law of universal gravitation.

During the 1970s, astronomers discovered something odd about the movement of stars in galaxies. Like the planets orbiting our sun, the stars should follow Newton's law of gravity, and travel ever more slowly the further out they are from the galactic centre. Yet beyond a certain distance, their speeds remained more or less constant - in flat contradiction of Newton's law.

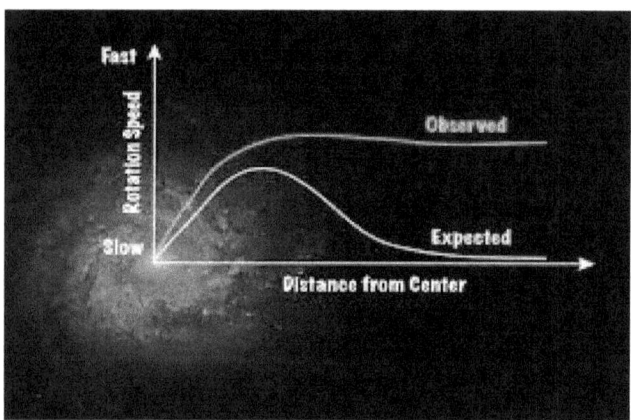

Astronomers quickly proposed a solution: that there are huge amounts of invisible "dark matter" lurking in and around galaxies, whose gravitational pull invisibly affects the stars. But Prof Milgrom had a more radical proposal: that there is something wrong with the law of gravity itself.

His calculations suggested that the anomalous motion of the stars could be explained if Newton's law breaks down for masses accelerating below a critical rate of around one ten-billionth of a metre per second per second.

For over 25 years Professor Mordehai Milgrom of the Weizmann Institute in Israel has been pursuing the possibility that both Newton and Einstein missed something when they devised their theories of this most ubiquitous of forces.

According to Einstein, mass warps the very fabric of space and time around it, rather like a cannonball sitting on a vast rubber sheet. This creates the illusion that objects moving past some mass are accelerated by a mysterious "force" emanating from it. In reality, they are just responding to the distortion of space and time - the effect of which is described in detail by Einstein's theory, and captured pretty well even by Newton's simple formula.

Since the early 1980s, Prof Milgrom has suspected there is another flaw in Newton's venerable formula - one which even Einstein failed to fix. And after decades of being ignored by the scientific establishment, there is mounting evidence that he is right.

Prof Milgrom's theory goes by the prosaic name of Modified Newtonian Dynamics or MOND, and is based the bizarre idea that Newton's law of gravity breaks down at low accelerations. And he means very low: around 100-billionth that generated by the Earth's gravity. Like Newton, Prof Milgrom was inspired by a simple observation - albeit a rather more esoteric one than the fall of an apple.

Invention

MOND was proposed by Mordehai Milgrom in 1983. The basic premise of MOND is that while Newton's laws have been extensively tested in high-acceleration environments, they have not been verified for objects with extremely low acceleration, such as stars in the outer parts of galaxies. Several independent observations point to the fact that the visible mass in galaxies and galaxy clusters is insufficient to account for their dynamics, when analysed using Newton's laws.

While Newton's Laws predict that stellar rotation velocities should decrease with distance from the galactic centre, Rubin and collaborators found instead that they remain almost constant, the rotation curves are said to be "flat". This observation necessitates at least one of the following:

1) There exists in galaxies large quantities of unseen matter which boosts the stars' velocities beyond what would be expected on the basis of the visible mass alone, or

2) Newton's Laws do not apply to galaxies.

The former leads to the dark matter hypothesis; the latter leads to MOND

In the disc galaxies most of the mass is at the centre of the galaxy, this means that if you want to calculate how a star moves far away from the centre it is a good approximation to only ask what is the gravitational pull that comes from the centre bulge of the galaxy.

Einstein taught us that gravity is really due to the curvature of space and time but in many cases it is still quantitatively incorrect to describe gravity as a force, this is known as the Newtonian limit and is a good approximation as long as the pull of gravity is weak and objects move much slower than the speed of light. It is a bad approximation for example close by the horizon of a black hole but it is a good approximation for the dynamics of galaxies that we are looking at here.

It is then not difficult to calculate the stable orbit of a star far away from the centre of a disc galaxy. For a star to remain on its orbit, the gravitational pull must be balanced by the centrifugal force,

$$\frac{mv^2}{R} = \frac{GMm}{R^2}$$

You can solve this equation for the velocity of the star and this will give you the velocity that is necessary for a star to remain on a stable orbit,

$$v = \sqrt{\frac{GM}{R}}$$

As you can see the velocity drops inversely with the square root of the distance to the centre. But this is not what we observe, what we observe instead is that the velocity continue to increase with distance from the galactic centre and then they become constant.

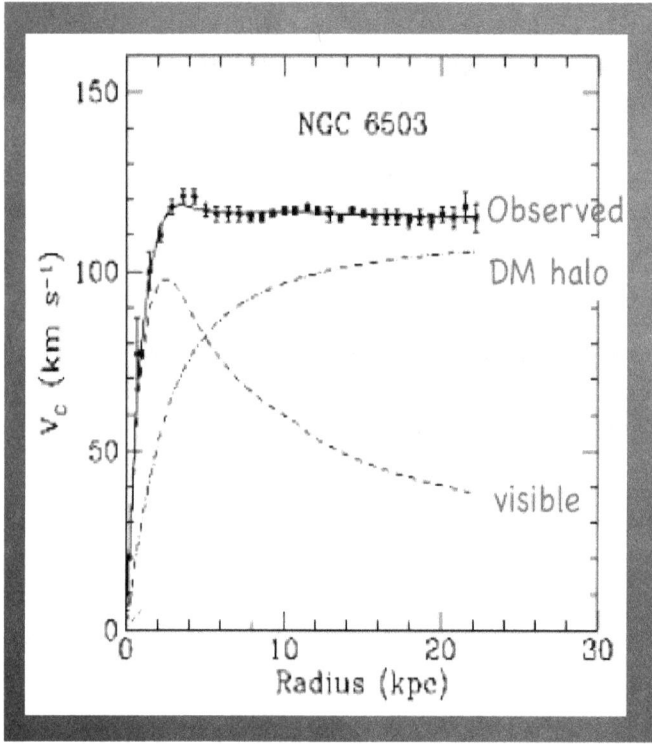

Galaxy data that show that Newtonian and Einstein gravity do not fit the observed speed of stars in orbits inside a galaxy such as NGC 6503

This is known as the flat rotation curve. This is not only the case for our own galaxy but it is the case for hundred of galaxies that have been observed. The curves don't always become perfectly constant sometimes they have rigorous lines but it is abundantly clear that these observations cannot be explained by the normal matter only.

Dark matter solves this problem by postulating that there is additional mass in galaxies distributed in a spherical halo. This has the effect of speeding up the stars because the gravitational pull is now stronger due to the mass from the dark matter halo. There is always a distribution of dark matter that will reproduce whatever velocity curve we observe.

In contrast to this, Modified Newtonian Dynamics (MOND) postulates that gravity works differently. In MOND, the gravitational potential is the logarithmic of the distance

$$\Phi = \left(\sqrt{GMa_o}\right) \ln\left(\frac{R}{GM}\right)$$

and not as normally the inverse of the distance

$$\Phi = \frac{-GM}{R}$$

In MOND the gravitational force is then the derivation of the potential that is, the inverse of the distance

$$F = \frac{\sqrt{GMa_o}}{R}$$

while normally it is the inverse of the square of the distance

$$F = \frac{GMm}{R^2}$$

If you put the modified gravitational force into the force balance equation as before

$$\frac{\sqrt{GMa_o}}{R} = \frac{v^2}{R}$$

you will see that the dependence on the distance cancels out and the velocity just becomes constant. Now of course you cannot just go and throw out the normal $\frac{1}{R^2}$ gravitational force law because we know that it works on the solar system. Therefore MOND postulates that the normal $\frac{1}{R^2}$ law crosses over into a $\frac{1}{R}$ law. This crossover happens not at a certain distance but it happens at a certain acceleration.

The New force law comes into play at low acceleration a_o, this acceleration where the crossover happens is a free parameter in MOND. You can determine the value of this pararmeter by just trying out which fits the data best. It turns out that the best fit value is closely related to the cosmological constant $a_o \approx \sqrt{\frac{\Lambda}{3}}$. In addition to demonstrating that rotation curves in MOND are flat, it provides a concrete relation between a galaxy's total baryonic mass (the sum of its mass in stars and gas) and its asymptotic rotation velocity. Observationally, this is known as the baryonic Tully Fisher relation and is found to conform quite closely to the MOND prediction.

Milgrom's law fully specifies the rotation curve of a galaxy given only the distribution of its baryonic mass. In particular, MOND predicts a far stronger correlation between features in the baryonic mass distribution and features in the rotation curve than does the dark matter hypothesis.

It predicts a specific relationship between the acceleration of a star at any radius from the centre of a galaxy and the amount of unseen (dark matter) mass within that radius that would be inferred in a Newtonian analysis. This is known as the "mass discrepancy-acceleration relation", and has been measured observationally.

One aspect of the MOND prediction is that the mass of the inferred dark matter go to zero when the stellar centripetal acceleration becomes greater than a_0, where MOND reverts to Newtonian mechanics. In dark matter hypothesis, it is a challenge to understand why this mass should correlate so closely with acceleration, and why there appears to be a critical acceleration above which dark matter is not required.

In MOND, all gravitationally bound objects with $a < a_0$ regardless of their origin – should exhibit a mass discrepancy when analysed using Newtonian mechanics, and should lie on the BTFR. Under the dark matter hypothesis, objects formed from baryonic material ejected during the merger or tidal interaction of two galaxies are expected to be devoid of dark matter and hence show no mass discrepancy. Three objects unambiguously identified as Tidal Dwarf Galaxies appear to have mass discrepancies in close agreement with the MOND prediction.

Recent work has shown that many of the dwarf galaxies around the Milky Way and Andromeda are located preferentially in a single plane and have correlated motions. This suggests that they may have formed during a close encounter with another galaxy and hence be Tidal Dwarf Galaxies. If so, the presence of mass discrepancies in these systems constitutes further evidence for MOND.

By itself, Milgrom's law is not a complete and self-contained physical theory, but rather an ad-hoc empirically motivated variant of one of the several equations that constitute classical mechanics. Its status within a coherent non-relativistic hypothesis of MOND is akin to Kepler's third law within Newtonian mechanics; it provides a succinct description of observational facts, but must itself be explained by more fundamental concepts situated within the underlying hypothesis.

The majority of astrophysicists and cosmologists accept dark matter as the explanation for galactic rotation curves, and are committed to a dark matter solution of the missing-mass problem.

MOND, by contrast, is actively studied by only a handful of researchers. The primary difference between supporters of ΛCDM and MOND is in the observations for which they demand a robust, quantitative explanation and those for which they are satisfied with a qualitative account, or are prepared to leave for future work.

This invisible and undetected matter removes any need to modify Newton's and Einstein's gravitational theories. Invoking dark matter is a less radical, less scary alternative for most physicists than inventing a new theory of gravity.

If dark matter is not detected and does not exist, then Einstein's and Newton's gravity theories must be modified. Can this be done successfully? Yes! My Quantum force equation can explain the astrophysical, astronomical and cosmological data without dark matter.

A New View on Gravity

Question1.

Galaxy rotation problem: Is dark matter responsible for differences in observed and theoretical speed of stars revolving around the centre of galaxies, or is it something else?

The galaxy rotation problem is the discrepancy between observed galaxy rotation curves and the theoretical prediction, assuming a centrally dominated mass associated with the observed luminous material. A solution to this conundrum is to hypothesize the existence of dark matter and to assume its distribution from the galaxy's center out to its halo. Although this seems to be a good approach towards the problem, it leaves many questions unanswered as we saw before in the preceding sections. The only way forward is to discover by theoretical means the origin of gravity. If we discover where gravity comes from, we will not only be able to solve the galaxy rotation problem but we will also be able to develop a quantum theory of gravity and later the theory of everything. Allow me in this section to take you on a journey towards the origin of gravity.

Origin of Gravity

The real origin of gravity is one of the most important, complex and substantially yet unsolved questions in Physics. The replacement of the Newtonian model of gravity with the Einstein's one given by General Relativity (GR) has only shifted the question without solving it. Within GR, gravity has two possible interpretations: a field one and a geometric one. According to the latter, that has become the prevalent one, gravity is due to the curvature of the space – time "tissue", represented as a "rubber sheet", due to the presence of a mass. Nevertheless, this is a purely mathematical description telling nothing about the physical mechanism starting the motion. In fact, even supposing the existence, in the neighbouring of a source mass, of a curved four dimensional manifold it doesn't explain why a second particle at rest should move towards the source mass.

As such, it invites attempts at derivation from a more fundamental set of underlying assumptions, and six such attempts are outlined in the standard reference book Gravitation, by Misner, Thorne, and Wheeler (MTW). ' Of the six approaches presented in MTW, perhaps the most far-reaching in its implications for an underlying model is one due to Sakharov; namely, that gravitation is not a fundamental interaction at all, but rather an induced effect brought about by changes in the quantum fluctuation energy of the vacuum when matter is present. 'In this view the attractive gravitational force is more akin to the induced van der Waals and Casimir forces, than to the fundamental Coulomb force. Although speculative when first introduced by Sakharov in 1967, this hypothesis has led to a rich and ongoing literature on quantum-fluctuation-induced gravity that continues to be of interest.

Many physicists believe that gravity, and space-time geometry are emergent. Also string theory and its related developments have given several indications in this direction.

In this section we will argue that the central notion needed to derive gravity is vacuum polarization. More precisely, the presence of matter in the vacuum is taken to constitute a kind of set of boundaries as in a generalized Casimir effect, and the question of how quantum fluctuations of the vacuum under these circumstances can lead to an action and metric that reproduce Einstein gravity will be addressed from several viewpoints. **We want to show that gravitation might be not a fundamental interaction but a byproduct of the electromagnetic interaction, precisely an electromagnetic phenomena induced by the presence of matter in the quantum vacuum (the quantum field that is present even in empty space).** Which means that, matter is not just there but is in the quantum vacuum, and therefore interacts with it, causing some kind of quantum fluctuation energy, that fluctuation is gravitation. In simple terms, a body immersed in quantum fields will interact with them causing gravity to manifest.

Changes in this vacuum when matter is displaced leads to a reaction force. Our aim is to show that this force, given certain reasonable assumptions, takes the form of gravity.

Vacuum polarization describes a process in which a background electromagnetic field produces virtual electron-positron pairs that change the distribution of charges and currents that generated the original electromagnetic field.

The effects of Vacuum polarization have been observed experimentally, for example in measurements of the lamb shift and anomalous magnetic dipole moment of the electron for a precise determination of the fine structure constant. These effects were calculated to first order in the coupling constant by R. serber and E.A Uehling in 1935.

The law of gravitation is derived from classical Casimir effect, which states that the vacuum through which particles move is not empty but consists of an indeterminate state of fluctuating fields and particles.

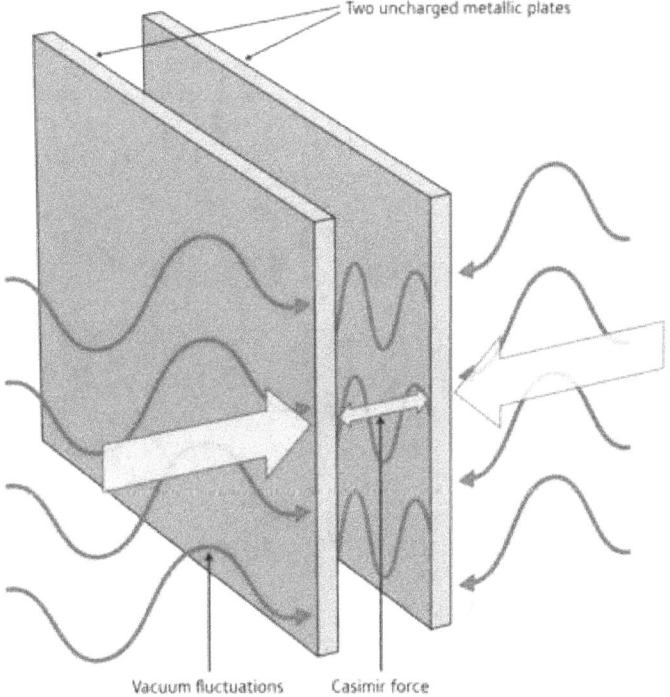

Even if you remove all the particles and radiation from a region of space, space still won't be empty. It will consist of virtual pairs of particles and antiparticles.

It is known that, when light propagates through an "empty" region, if space is perfectly empty, it should move through that space unimpeded, without bending, slowing or breaking into multiple wavelengths. Applying an external magnetic field doesn't change this, as photons, with their oscillatory electric and magnetic fields don't bend in a magnetic field. Even when your space is filled with particle antiparticle pairs, this effect doesn't change. But if you apply a **strong magnetic field** to a space filled with particle antiparticle pairs, suddenly a real, observable effect arises.

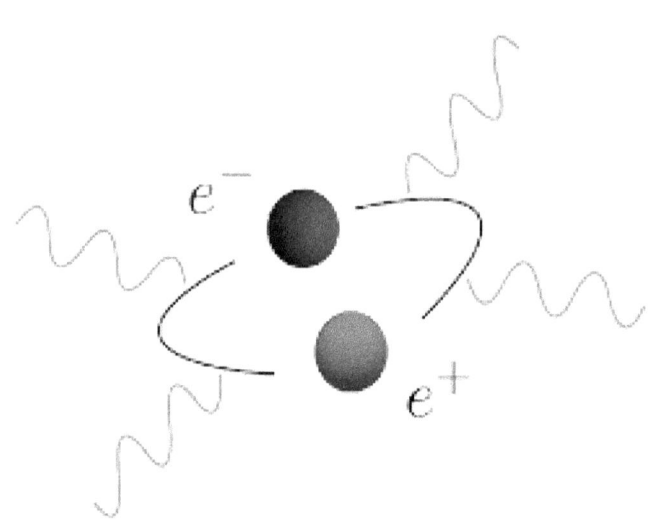

A virtual particle is a transient quantum fluctuation that exhibits some of the characteristics of an ordinary particle, while having its existence limited by the uncertainty principle. The longer the virtual particle exists, the closer its characteristics come to those of ordinary particles. They are important in the physics of many processes, including particle scattering and Casimir forces.

In 2015 theoretical physicist James Quach, suggested a way to detect gravitons by taking advantage of their quantum nature. Quantum mechanics suggests the universe is inherently fuzzy for instance, one can never absolutely know a particle's position and momentum at the same time. One consequence of this uncertainty is that a vacuum is never completely empty, but instead buzzes with a "quantum foam" of so-called virtual particles that constantly pop in and out of existence. These ghostly entities may be any kind of quanta, including gravitons.

Decades ago, scientists found that virtual particles can generate detectable forces. For example, the Casimir effect is the attraction or repulsion seen between two mirrors placed close together in vacuum. These reflective surfaces move due to the force generated by virtual photons winking in and out of existence. Previous research suggested that superconductors might reflect gravitons more strongly than normal matter, so Quach calculated that looking for interactions between two thin superconducting sheets in vacuum could reveal a gravitational Casimir effect. The resulting force could be roughly 10 times stronger than that expected from the standard virtual-photon-based Casimir effect.

In quantum field theory, even classical forces-such as the electromagnetic repulsion or attraction between two charges-can be thought of as due to the exchange of many virtual photons between the charges. Virtual photons are the exchange particle for the electromagnetic interaction.

From the energy time uncertainty principle, we deduce the Casimir force in this way,

$$\Delta E \Delta t \sim \hbar$$

Considering two electrons separated by a distance R in a vacuum, the work done by a photon to move from one charge to the other is equivalent to the product of the force F applied and the distance R through which it moves

$$\Delta E = F \Delta R$$

Therefore the force carried by the virtual photon (messenger particles) between the two charges is

$$F = \frac{\hbar}{\Delta R \Delta t}$$

But since the virtual photons move at a constant speed of light c in the vacuum between the electrons separated by a distance ΔR, inserting

$$c = \frac{\Delta R}{\Delta t}$$

The Casimir force is given as,

$$F = \frac{\hbar c}{R^2}$$

We notice that the force deduced above is similar to the gravitational force because it falls off as the inverse of the distance squared.

An electron feels a force like gravity due to the changes in the vacuum (exchange of gravitons) brought about by the presence of a strong magnetic field.

Recently, Norte and his colleagues developed a microchip to perform this experiment. This chip held two microscopic aluminum-coated plates that were cooled almost to absolute zero so that they became superconducting. One plate was attached to a movable mirror, and a laser was fired at that mirror. If the plates moved because of a gravitational Casimir effect, the frequency of light reflecting off the mirror would measurably shift. However the scientists failed to see any gravitational Casimir effect. This null result does not necessarily rule out the existence of gravitons and thus gravity's quantum nature. Rather, it may simply mean that gravitons do not interact with superconductors as strongly as prior work estimated, says quantum physicist and Nobel laureate Frank Wilczek of the Massachusetts Institute of Technology, who did not participate in this study and was unsurprised by its null results. Even so, Quach says, this "was a courageous attempt to detect gravitons."

Although Norte's microchip did not discover whether gravity is quantum, other scientists are pursuing a variety of approaches to find gravitational quantum effects. For example, in 2017 two independent studies suggested that if gravity is quantum it could generate a link known as "entanglement" between particles, so that one particle instantaneously influences another no matter where either is located in the cosmos. A table top experiment using laser beams and microscopic diamonds might help search for such gravity-based entanglement. The crystals would be kept in a vacuum to avoid collisions with atoms, so they would interact with one another through gravity alone. Scientists would let these diamonds fall at the same time, and if gravity is quantum the gravitational pull each crystal exerts on the other could entangle them together.

The researchers would seek out entanglement by shining lasers into each diamond's heart after the drop. If particles in the crystals' centers spin one way, they would fluoresce, but they would not if they spin the other way. If the spins in both crystals are in sync more often than chance would predict, this would suggest entanglement. "Experimentalists all over the world are curious to take the challenge up," says quantum gravity researcher Anupam Mazumdar of the University of Groningen in the Netherlands, co-author of one of the entanglement studies.

To prove that gravity is Casimir, we propose that the virtual photons exchanged between electrons take the force of electromagnetic waves or radiations propagating in the vacuum between the electrons with a changing electric and magnetic field, such that their speed depends on the electric field E and magnetic B as,

$$c = \frac{E}{B}$$

The force felt by an electron due to vacuum polarization then becomes,

$$F = \frac{\hbar E}{BR^2} \quad (1)$$

It was previously deduced (see my book "quantum gravity in a Nutshell1") that the characteristic value of the electric field built from the electron mass is

$$E = \frac{m^2 ec^2}{4\pi\varepsilon_o \hbar^2} = 9.667 \times 10^{15} \text{N/C}$$

Where m is the mass of the electrons, e is the elementary charge and ε_o is the permittivity of the vacuum between the electrons

This electric field can be observed near Neutron stars or magnetars.

$$F = \frac{m^2 ec^2}{4\pi\varepsilon_o \hbar BR^2}$$

Suppose we apply a strong external Magnetic field to the vacuum of

$$B = \frac{ec^2}{4\pi\varepsilon_0 G\hbar} = 1.8423 \times 10^{48} T$$

Note: The above given magnetic field was previously derived in my book "quantum gravity in a nutshell1"

We obtain a familiar law between the two electrons

$$F = G\frac{m^2}{r^2}$$

We have recovered Newton's law of gravitation, practically form first principles. Following the above derivation carefully, it implies that gravity is a force resulting from the quantum vacuum polarizations due to an existence of an external strong magnetic field. This is proof that gravity is indeed a quantum force which can be explained from the Casimir effect.

The effects of the given derivation can be experimentally verified near Neutron stars and Magnetars with strong magnetic fields. When the given values of the electric and magnetic fields are observed with high tech experiments then it will prove without doubt that the force felt by the electron in the vaccum is indeed a quantum force.

The above given theoretical model can be verified experimentally from the effect known as Vacuum Birefringence, occurring when charged particles, get yanked in opposite directions by strong magnetic field lines.

The effect of this vacuum birefringence gets stronger very quickly as the magnetic field strength increases, as the square of the field strength. Even though the effect is small, we have place in the universe where the magnetic field strength get large enough to make these effects relevant. This place is the Neutron star and Magnetars with strong magnetic fields.

The outer 10% of a neutron star consists mostly of protons, light nuclei, and electrons, which can stably exist without being crushed at the neutron star's surface.

Neutron stars rotate extremely rapidly, frequently in excess of the 10% the speed of light, meaning that these charged particles on the outskirts of the neutron star are always in motion, which necessitates the production of both electric currents and induced magnetic fields. These are the fields we should be looking for if we want to observe vacuum birefringence, and its effects on the polarization of light.

All the light that's emitted must pass through the strong magnetic field around the neutron star on its way to our eyes, telescope and detectors, if the magnetized space that it passes through exhibits the expected vacuum birefringence effect, that light should all be polarized, with a common direction of polarization for all the photons.

In summary, Newton's law of gravitation naturally arise in a theory in which space emerges through a zero- point fluctuation of the quantum vacuum. Gravity is identified with a casimir force caused by quantum vacuum fluctuations due to the presence of material bodies in it or the distortion of the vacuum through its interaction with mass. A relativistic generalization of the presented arguments directly leads to the Einstein equations. When space is emergent even Newton's law of inertia needs to be explained. The equivalence principle suggests that it is actually the law of inertia whose origin is casimir.

Observational Evidence

1. The Tully-Fisher relation

Equation (1) above can be written in its simplest form

$$F = \frac{\alpha \hbar c}{R^2}$$

Where $\alpha = \frac{B_o}{B}$, $(B_o = Ec)$ is the coupling constant which determines the strength of the gravitational force in an interaction, $\hbar$ is the reduced Planck constant, c is the constant speed of light.

But since the energy density or pressure of matter confined in a given region of "empty" space is force per unit area or energy per unit volume,

$$\rho = \frac{F}{4\pi R^2}$$

Then the reaction force felt by the space in which a particle rests or moves is given as,

$$F = \sqrt{4\pi\alpha\hbar c\rho} \qquad (2)$$

The above given force is here termed as a "quantum force" of nature.

Applying this force to the entire galaxy, Now suppose our boundary is not infinitely extended, but forms a closed surface. More specifically, let us assume it is a sphere with already emerged space on the outside.

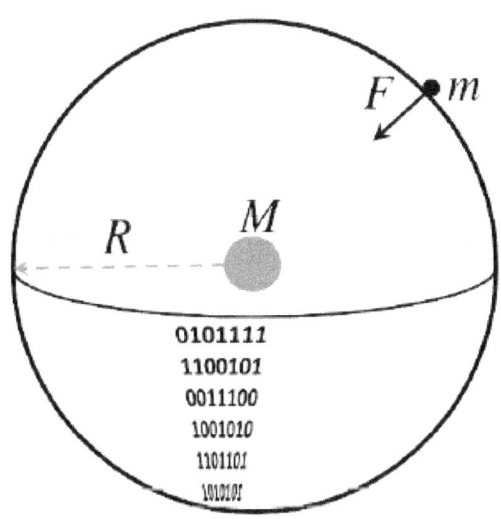

A particle with mass m near a spherical holographic screen. The magnetic field is evenly distributed over the occupied bits, and is equivalent to the mass M that would emerge in the part of space surrounded by the screen.

Let the surface of the boundary be evenly distributed with a positive elementary charge +e creating in turn a negative charge on the inside.

The key statement is simply that we need to have vacuum energy density in order to have a force between the two masses. Since we want to understand the origin of gravity, we need to know where the vacuum energy density comes from.

One can think about the boundary as a storage device for information. Assuming that the holographic principle holds, the maximal storage space, or total number of bits, is proportional to the internal Electric Field created by the masses. In fact, in a theory of emergent space this is how magnetic field may be defined: each fundamental bit occupies by definition one unit cell.

Let us denote the number of used bits by α. It is natural to assume that this number will be proportional to the Electric field. So we write as before,

$$\alpha = \frac{B_o}{B} = \frac{\left(\frac{Mmec}{4\pi\varepsilon_o\hbar^2}\right)}{\left(\frac{ec^2}{4\pi\varepsilon_o G\hbar}\right)} = \frac{GMm}{\hbar c}$$

Note here that we have just given the origin of the gravitational coupling constant.

Where we introduced a new constant G. Eventually this constant is going to be identified with Newton's constant, of course. But since we have not assumed anything yet about the existence of a gravitational force, one can simply regard this equation as the definition of G. So, the only assumption made here is that the number of bits is proportional to the characteristic Electric field from the mass. Nothing more.

Suppose there is a total vacuum energy density ρ present in the system. The vacuum energy density is determined by the equipartition rule

$$\rho = \frac{F}{A}$$

Finally one inserts

$$A = 4\pi R^2$$

The energy density is then defined as the force on particles accelerating in the vacuum per unit area,

$$\rho = \frac{F_o}{4\pi R^2}$$

From Newton's second law, the force on the particle is related to the acceleration by,

$$F_o = ma_o$$

In contrast to Dark matter, our quantum force postulates that gravity works differently. The gravitational force is the inverse of the distance

$$F = \frac{1}{R}\sqrt{\alpha \hbar c m a_o}$$

Applying this to an object of mass m in circular orbit around a point mass M (a crude approximation for a star in the outer regions of a galaxy), we find:

$$\frac{mv^2}{R} = \frac{1}{R}\sqrt{\alpha \hbar c m a_o}$$

you will see that the dependence on the distance cancels out and the velocity just becomes constant. Finally one inserts the number of bits/gravitational coupling constant

$$\alpha = \frac{GMm}{\hbar c}$$

and one obtains the familiar relation

$$v^4 = a_o GM$$

We have recovered the Tully-Fisher relation, practically from first principles! It is one of the best fit predictions for MOND " The relation between asymptotic velocity and the mass of the galaxy is an absolute one" (Milgrom 1983).

This is given by, $V^4 = a_o GM$, where $a_o = 1.2 \times 10^{-10} ms^{-2}$. It is this behavior that gives rise to asymptotically flat rotation curves and the Tully-Fisher relation (Tully & Fisher 1977) without invoking dark matter.

The quantum force is an alternative to the hypothesis of dark matter in terms of explaining why galaxies do not appear to obey the currently understood laws of physics. It is an alternative to Entropic gravity, MOND, Quantum gravity and General relativity. Now of course you cannot just go and throw out the normal

$$\frac{1}{r^2}$$

gravitational force law because we know that it works on the solar system. Therefore the normal $\frac{1}{r^2}$ law crosses over into a $\frac{1}{r}$ law. This crossover happens not at a certain distance but it happens at certain acceleration. The New force law comes into play at low acceleration $a_o = 1.2 \times 10^{-10} ms^{-2}$, this acceleration where the crossover happens is a free parameter in MOND. You can determine the value of this pararmeter by just trying out which fits the data best. It turns out that the best fit value is closely related to the cosmological constant

$$a_o \approx \sqrt{\frac{\Lambda}{3}}$$

We anticipate that this quantum force will modify how stars collapse and the nature of black holes.

2. The Cosmological constant

The vacuum energy of free space has been estimated to be 10^{-9} joules per cubic meter. However, in both quantum electrodynamics (QED) and stochastic electrodynamics (SED), consistency with the principle of Lorentz covariance and with the magnitude of the Planck constant suggest a much larger value of 10^{113} joules per cubic meter. This huge discrepancy is known as the cosmological constant problem.

The question is; why does the Zero-point energy of the vacuum not cause a large Cosmological constant? What cancels it out?

Attempts to find a way to measure or to calculate the value of the vacuum energy density and the cosmological constant have all either failed or produced results incompatible with observations or other confirmed theoretical results. Some of those results are theoretically implausible because of certain unrealistic assumptions on which the calculation model is based. And some theoretical results are in conflict with observations, the conflict itself being caused by certain questionable hypotheses on which the theory is based. And the best experimental (Casmir effect) evidence is based on the measurement of the difference of energy density within and outside of the measuring apparatus, thus preventing in principle any numerical assessment of the actual energy density.

Below I calculate both the experimental observed and theoretical vacuum energy density values from first principles and thereafter give a reason for the cause of the small cosmological constant value.

The Experimental observed value

Consider a Planck particle of mass $m_p = 2.18 \times 10^{-8} kg$ accelerating at a low gravitational acceleration limit in MOND of $a_o = 1.2 \times 10^{-10} ms^{-2}$. From Newton second law, the force that keeps the Planck particle in motion is $F = m_p a_o$. From equation (2), the vacuum energy density is then given as,

$$\rho = \frac{M_p^2 a_o^2}{4\pi \alpha \hbar c}$$

Let α be the electromagnetic coupling/ fine structure constant defined by the value $\alpha = \frac{1}{137.036}$. Putting this into account, we get the vacuum energy density value in agreement with the Planck collaboration 2018 values as,

$$\rho = 2.358 \times 10^{-9} \frac{J}{m^3}$$

This is the main formula and central result of the cosmological constant, since it allows one to make a direct comparison with observations.

The theoretical calculated value

Consider a Planck particle of mass $m_p = 2.18 \times 10^{-8} kg$ accelerating at a high gravitational acceleration limit in Quantum gravity of $a_{max} = 5.5608 \times 10^{51} ms^{-2}$. From Newton second law, the force that keeps the Planck particle in motion is $F = m_p a_{max}$. From equation (2), the vacuum energy density is then given as,

$$\rho = \frac{M_p^2 a_{max}^2}{4\pi\alpha\hbar c}$$

Let α be the electromagnetic coupling/ fine structure constant defined by the value $\alpha = \frac{1}{137.036}$. Putting this into account, we get the agreement with the theoretical calculated values as,

$$\rho = 5.0634 \times 10^{114} \frac{J}{m^3}$$

Resolution of the cosmological constant problem

The cosmological constant Λ is a dimensionful parameter with units of $(length)^{-2}$. From the point of view of classical general relativity, there is no preferred choice for what the length scale defined by Λ might be. Particle physics, however, brings a different perspective to the question. Einstein introduced a cosmological constant into his equations for General Relativity. This term acts to counteract the gravitational pull of matter, and so it has been described as an anti-gravity effect. The cosmological constant turns out to be a measure of the energy density but no one has ever calculated the cosmological constant with confidence. The cosmological constant is related to the vacuum energy density by the Friedmann relationship as,

$$\Lambda = \frac{8\pi G}{c^4}\rho$$

When we substitute for (2), we get the value of the cosmological constant as,

$$\Lambda = \frac{2Gm_p{}^2 a_o{}^2}{\alpha\hbar c^5} = 4.88 \times 10^{-52} m^{-2}$$

This value of the cosmological constant is also in agreement with the Planck collaboration 2018 values.

Arranging the above given formula into its original physical constants we get

$$\Lambda = \frac{8\pi\varepsilon_o \hbar}{c^3}\left(\frac{a_o}{e}\right)^2 = constant \times a_o{}^2$$

Where e is the elementary charge and ε_o is the permitivity of free space

Therefore the Zero-point energy of the vacuum doesn't cause a large Cosmological constant because in the observable universe at a Planck scale $m_p{}^2$, Newton and Einstein gravity are modified at a low acceleration limit $a_o = 1.2 \times 10^{-10} ms^{-2}$. It is this acceleration due to the quantum vacuum fluctuations according to equation2 that cancels out the large cosmological constant. Finally the reason for this small value of the cosmological constant is because gravity is a result of the quantum vacuum fluctuations that lead particles with a Planck mass to move at low accelerations in a form of what we call space time.

6. The Simple Link between Quantum Mechanics and Gravity

Today we are blessed with two extraordinarily successful theories of physics. The first is the General theory of relativity, which describe the large scale behavior of matter in a curved space time. This theory is the basis for the standard model of big bang cosmology. The discovery of gravitational waves at LIGO observatory in the US (and then Virgo, in Italy) is only the most recent of this theory's many triumphs.

The second is quantum mechanics. This theory describes the properties and behavior of matter and radiation at their smallest scales. It is the basis for the standard model of particle physics, which builds up all the visible constituents of the universe out of collections of quarks, electrons and force-carrying particles such as photons. The discovery of the Higgs boson at CERN in Geneva is only the most recent of this theory's many truimphs.

But, while they are both highly successful, those two structures leave a lot of important questions unanswered. They are also based on two different interpretations of space and time, and are therefore fundamentally incompatible. We have two descriptions but, as far as we know, we've only ever had one universe. What we need is a quantum theory of gravity.

One of the problems facing physicists is to link Gravity with quantum mechanics in the study of the quantum effects of Black holes. Within the past few years there have been developments that give rise to the hope that before too long we shall have a fully consistent quantum theory of gravity, one that will agree with general relativity for macroscopic objects and will, one hopes, be free of the mathematical infinities that have long bedeviled other quantum field theories. Such developments include; Loop Quantum Gravity and String theory. These developments have to do with certain recently discovered quantum effects associated with black holes, which provide a remarkable connection between black holes and the laws of thermodynamics.

One of the first attempts towards the unification of quantum mechanics with gravity was made in part by Stephen Hawking and Bekenstein in the late 70s. The black hole creates and emits particles and radiation just as if it were an ordinary hot body with a temperature that is proportional to the surface gravity and inversely proportional to the mass. This made Bekenstein's suggestion that a black hole had a finite entropy fully consistent, since it implied that a black hole could be in thermal equilibrium at some finite temperature other than zero.

Hawking beautiful result raises a number of questions. First, in Hawking's derivation the quantum properties of gravity are neglected. Are these going to affect the result?

Second, we understand macroscopical entropy in statistical mechanical terms as an effect of the microscopical degrees of freedom. What are the microscopical degrees of freedom responsible for the entropy? Can we derive the Bekenstein entropy from first principles? Because of the appearance of $\hbar$ in the entropy formula, it is clear that the answer to these questions has since become a standard benchmark against which a quantum theory of gravity can be tested.

A definitive resolution of the quantum gravity problem will require deriving the Bekenstein-Hawking area-entropy law from first principles without the introduction of adhoc principles.

$$S = \frac{Ac^3 k}{4\hbar G}$$

where A is the surface area of the Schwarzschild black hole, c is a constant speed of light, k the Boltzmann constant, $\hbar$ the reduced Planck constant and G is the Newton's gravitational constant.

Attempts towards this were done in the early 70s by Hawking who proved that a black hole emits thermal radiation with a temperature

$$T = \frac{\hbar c^3}{8\pi G k}$$

This book presents a simple universal explanation of Black hole thermodynamics in a somewhat different form than that given by Loop Quantum Gravity (LQG), String theory and Hawking radiation theory. The major result of the book is the derivation of the temperature and entropy of a black from first principles with a well defined calculation where no infinities appear.

Quantizing Gravity

Gravity is the weakest of the four fundamental forces of physics, approximately 10^{38} times weaker than the strong force, 10^{36} times weaker than the electromagnetic force and 10^{29} times weaker than the weak force. As a consequence, it has no significant influence at the level of subatomic particles. This becomes a problem in reconciling gravity with quantum mechanics. To unit gravity with the other three forces of physics, we propose that gravity be quantized in the following form

$$F_G = \alpha F_I \qquad (1)$$

Where F_G is the gravitational force, α is the quantum number and F_I is one of the other three forces of nature.

In case one wants to determine the strength of the gravitational force in comparison to the other three forces, then one sets α as the gravitational coupling constant which determines the strength of the force

$$\alpha = \frac{F_G}{F_I}$$

Although gravity is one of the weakest forces known in the universe, it however becomes stronger near space time singularities and from Einstein general relativity, a particle at the event horizon of a Black hole (that is at a Schwarzichild radius $R = 2GM/c^2$ from the center of a Black hole) feels a strong force given as

$$F_G = \frac{GM^2}{\pi R^2} = \frac{c^4}{4\pi G}$$

Note from above, the known gravitational force law is inversely proportional to the area of the event horizon of a Black hole not it's radius.

If a black hole was an oscillating particle like an electron, then according to quantum mechanics (from the uncertainty principle), its position will be determined by hitting it with a particle the same size as its mass.

$$x = \frac{\hbar}{2Mc}$$

Then the gravitational coupling constant becomes the ratio of the Schwarzschild radius to the position x

$$\alpha = \frac{R}{x} = \frac{4GM^2}{\hbar c}$$

Inserting in (1) we have a quantum mechanical description of gravity in the form

$$F_G x = R F_I$$

Then the work done by one of the three forces to move a particle from the center of a Black hole (in the form of radiation) to the event horizon through a Schwarzschild radius R against the strong gravitational force is then calculated to be

$$F_I R = \frac{\hbar c^3}{8\pi GM}$$

By the equipartition rule, the thermal energy of a Heat bath of a Black hole at temperature T is hereby equal to the work done by a force F_I to pull a particle from the black hole.

$$kT = F_I R = \frac{\hbar c^3}{8\pi GM}$$

Where k is the Boltzmann constant

Finally we get a familiar temperature

$$T = \frac{\hbar c^3}{8\pi GMk} \qquad (2)$$

We have therefore recovered Hawking temperature from first principles. This means that, the black hole creates and emits particles and radiation just as if it were an ordinary hot body with a temperature that is proportional to the surface gravity and inversely proportional to the mass.

When a Black hole evaporates or emits radiations, it losses mass in time t, the intensity of the emitted thermal or electromagnetic radiations is given by

$$I = \frac{F_I^2}{2\pi\hbar}$$

Where the intensity by definition is power P per unit area A of the event horizon of a black hole

$$\frac{P}{A} = \frac{F_I^2}{2\pi\hbar}$$

The power is the energy E lost by a black hole in time t (The time taken by a Black hole to dissipate)

$$t = \frac{Mc}{F_I}$$

From which the total energy lost by a Black hole is given by

$$E = \frac{AMcF_I}{2\pi\hbar}$$

But since from (1),
$$F_G = \alpha F_I$$

Then

$$E = \frac{AMcF_G}{2\pi\alpha\hbar}$$

Inserting $F_G = \frac{c^4}{4\pi G}$ and $\alpha = \frac{4GM^2}{\hbar c}$, we get

$$E = \frac{Ac^6}{32\pi G^2 M}$$

We have recovered the Frodden-Ghosh-Perez Energy.

Hawking knew that if the horizon area were an actual entropy, black holes would have to radiate. When heat is added to a thermal system, the change in entropy is the increase in thermal energy divided by temperature:

$$S = \frac{E}{T}$$

From (2), the temperature of a black hole is given by,

$$T = \frac{\hbar c^3}{8\pi GMk}$$

Therefore the Entropy of a Black hole will be given by,

$$S = \frac{\left(\frac{Ac^6}{32\pi G^2 M}\right)}{\left(\frac{\hbar c^3}{8\pi GMk}\right)} = \frac{Akc^3}{4G\hbar}$$

This implies that the energy radiated by a black hole goes into increasing its event horizon area and an increase in area will automatically lead to an increase in entropy as explained by the above given area entropy law. As far as this book is concerned there is no other theory from which such a calculation can proceed. Hence the book is the only one from which a detailed quantum theory of gravity precedes and where the result of the Bekenstein-Hawking area entropy law can be achieved.

7. Resolution of the Proton Radius Puzzle

Today the proton radius is measured via three methods that is, the spectroscopy, nuclear scattering and muonic hydrogen (2010 experiment) methods.

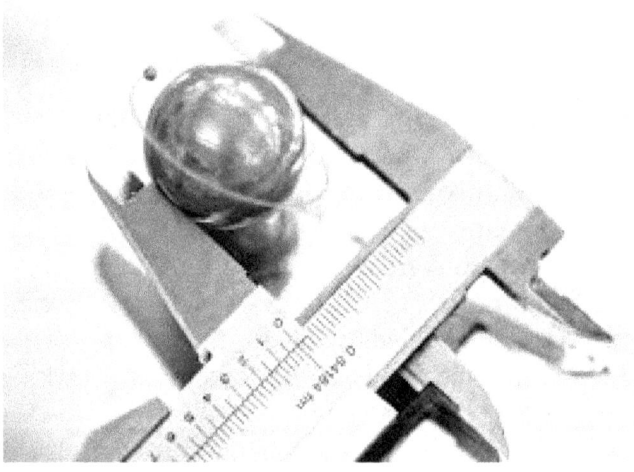

The spectroscopy method uses the energy levels of electrons orbiting the nucleus. This method produces a proton radius of about 8.768×10^{-16} m, with approximately 1% relative uncertainty.

The nuclear method is similar to Rutherford's scattering experiments that established the existence of the nucleus. Small particles such as electrons can be fired at a proton, and by measuring how the electrons are scattered, the size of the proton can be inferred. Consistent with the spectroscopy method, this produces a proton radius of about 8.775×10^{-16} m.

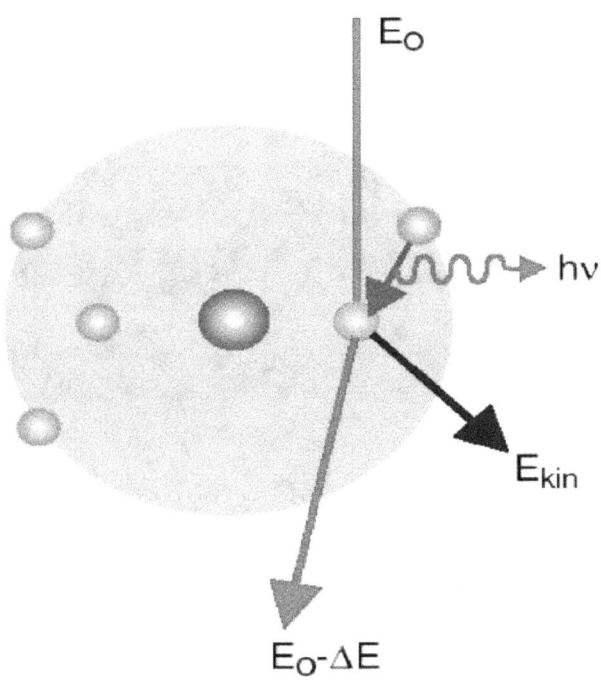

The muonic hydrogen 2010 method by Pohl et al. is similar to the spectroscopy method. However, the much higher mass of a muon causes it to orbit 207 times closer than an electron to the hydrogen nucleus, where it is consequently much more sensitive to the size of the proton. The resulting radius was recorded as 8.42×10^{-16} m. This newly measured radius is 4% smaller than the prior measurements, which were believed to be accurate within 1%.

The discrepancy between the measured values of the proton radius by the methods given above is what is called the proton radius puzzle and the discrepancy might be due to new physics, or the explanation may be an ordinary physics effect that has been missed.

In what follows, I deduce the radius of the proton from first principles using a new approach.

Consider two protons of mass m_{pro} at a distance R apart in an **antiprotonic hydroden atom**. Let the protons be circular with area A.

To measure the radius of the proton, we consider a **protonium**, also known as **antiprotonic hydroden**, in which a proton and an antiproton orbit each other.

According to the Heisenberg uncertainty principle, the space between the protons is not empty. Therefore the calculated energy density contained in the vacuum between the protons is then related to the Casimir force as

$$\rho_{cal} = \frac{F_c}{A} = \frac{2\hbar c}{AR^2} \qquad (1)$$

Where A is the area of the proton, $\hbar$ is the reduced Planck constant and c is the constant speed of light.

However, from general assumptions not given here, the observed/experimental vacuum energy density is proportional to the square of the force in any given interaction as,

$$\rho_{obs} = \frac{F^2}{8\pi\alpha\hbar c} \qquad (2)$$

Where α is the coupling constant (which determines the strength of the force in any given interaction)

At a point where the calculated vacuum energy density due to the Casimir effect is equal to the observed vacuum energy density due to the known fundamental force between the protons, the proton will move in a circular orbit around the other proton. The force of attraction between the two protons will then be deduced (by equating (1) to (2))

$$F = \frac{4\hbar c}{R}\sqrt{\frac{\pi\alpha}{A}} \qquad (3)$$

As the proton orbits closer to the other proton, its speed increases until it comes into contact with the other proton. Suppose the proton is moving at a constant speed of light c in its orbit, then the centripetal force that keeps it in orbit is equal to the force of attraction between them.

$$\frac{m_{pro}c^2}{R} = \frac{4\hbar c}{R}\sqrt{\frac{\pi\alpha}{A}}$$

Arranging and cancelling like terms, we get the formula for the area of the proton as

$$A = \frac{16\pi\alpha\hbar^2}{m_{pro}^2 c^2}$$

Inserting, $A = \pi r_{pro}^2$, where r_{pro} is the proton radius

$$r_{pro} = \frac{4\hbar}{m_{pro} c}\sqrt{\alpha}$$

The force required to separate two protons apart is indeed the strong nuclear force whose strength is determined by the strong coupling constant $\alpha = 1$, from which we calculate the proton radius as,

$$r_{pro} = \frac{4\hbar}{m_{pro} c} = 8.4075 \times 10^{-16} m$$

This value is very close to the 2010 Pohl result of an experiment relying on muonic hydrogen, practically from first principles! See Paul Scherrer institute in Switzerland (CREMA-Charge Radius Experiment with Muonic Atoms) vol466/8 july2010/doi:10.1038/nature09250

8. On the Irreducible Anomaly in Einstein Deflection Angle

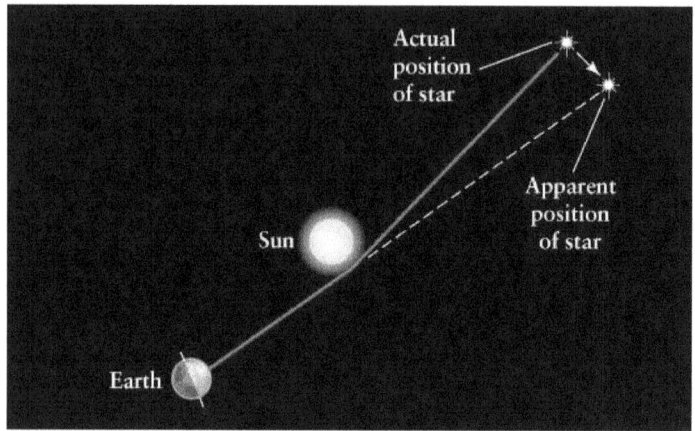

The Newtonian Approach

During Newton's time, it was believed that light was made up of particles moving at a varying speed.

To prove why light bends near the Sun's surface Newton had to assume that these particles had mass. For example he considered a Sun with mass M, where a particle of light with mass m from a distant star past the Sun, had to bend near the Sun's surface due to the gravitational force of attraction acting on the particle of light. Because of this, the observer at the earth's surface never saw the actual position of the star but rather the apparent position of the star at an angle θ from its original position. Newton assumed that, the particle of light falling freely in the gravitational field of the Sun gained kinetic energy,

$$E_k = \frac{mv^2}{2}$$

Where, v was the speed of the particle of light. The potential gravitational energy that was gained by the particle was given by,

$$V_k = \frac{GMm}{R}$$

Where R was the radius of the Sun from its centre to the point where light curved. Newton assumed that deflection angle was actually the ratio of the gravitational potential to the kinetic energy of the particle of light,

$$\theta = \frac{V_r}{E_k}$$

$$\theta = \frac{2GM}{v^2 R}$$

During Newton's time, the speed of the photon (a particle of light) was not known but, today we know this value to a much greater accuracy, thanks to Maxwell and Einstein. All the parameters, from the Sun's mass to the speed of light are known to high accuracy, therefore the Newtonian deflection angle is now known to be,

$$\theta_N = \frac{2GM}{c^2 R} = 0.875 arcsec$$

The problem with the Newtonian approach is this; we now know that photons of light are massless and move at a constant speed of light c and, the Newtonian deflection angle value is not in agreement with the observations. Therefore Newton's original calculation was flawed and required another explanation.

The Einstein Approach

In the Einstein approach it was found out that, the particles of light were called photons and that these particles where massles moving at a constant speed of light $c = 3 \times 10^8 m/s$. Einstein's theory proposes that gravity is not an actual force, but is instead a geometric distortion of spacetime not predicted by ordinary Newtonian physics. The more mass you have to produce the gravity in a body the more distortion you get, this distortion changes the trajectories of objects moving through space, and even the paths of light rays, as they pass close-by the massive body. Even so, this effect is very feeble for an object as massive as our own sun, so it takes enormous care to even detect that it is occurring.

The Einstein deflection angle was twice the Newton's angle of deflection $\theta = 2\theta_N$, but there is no any account in literature where it shows the derivation of this deflection angle from Einstein field equations, which means that, Einstein came up with a formula similar to the Newtonian deflection angle formula given as,

$$\theta_E = \frac{4GM}{c^2 R} = 1.75 arcsec$$

Instead of the number, **2** in the Newton formula, we have a **4** and the varying speed of light in the Newton approach is replaced by a constant speed of light c.

The Einstein value was determined by observation through observing the solar eclipse. Although they say it agrees with experiment, we know that this is not true. It has long been suspected that the deflection of light in the vicinity of the sun exceeds the general relativistic predicted value of 1.75". An example of this, is the Erwin Finlay Freundlich 1929 solar eclipse expedition which produced a value of 2.24" larger than the general relativistic value. It is expected that once the reason for the deviation in the deflection angle has been found, it will disprove Einstein's imaginations for the curvature of space time.

It's almost hundred years since Sir Arthur Eddington experimentally proved Einstein's general relativity theory right. Since then, there has never been any competing theory that would prove Einstein wrong save for Loop quantum gravity and string theory. The fact that starlight is bent at the surface of the gravitating body by a deflection angle of 1.75" imposes a bound on the theoretical justification of gravity. Calculating an angle below or above 1.75" will be an upheaval in the founding blocks of physics. Erwin Finlay Freundlich was one of those people who stood out of the ordinary in 1929 when he published results with a larger angle of deflection than Eddington's. An account on Freundlich 1929 expedition has been clearly given in Robert J.Trumpler and Klaus Hentschel papers as stated below;

"Among the various expeditions sent out to observe the total solar eclipse of May 9, 1929, that of the Potsdam Observatory (Einstein Stiftung) seems to be the only one which obtained photographs suitable for determining the light deflection in the Sun's gravitational field. Two instruments were used, but so far only the results of the larger one, a 28-foot horizontal camera combined with a coelostat, have been published. The three observers, Freundlich, von Klüber, and von Brunn, claim that these observations (four plates containing from seventeen to eighteen star images each) lead to a value of 2.24" for the deflection of a light ray grazing the Sun's edge; a figure that deviates considerably from the results of the 1922 eclipse, and which is in contradiction to Einstein's generalized theory of relativity".

The irreducible anomaly in the observations of the deflection of light by the sun has been known to exist since the birth of Einstein General relativity theory. For example, in a 1959 classical review by A.A.Mikhailov, it concludes that observations yield instead of a general relativistic prediction of 1.75arcsec at the limb of the sun the simple mean value of 2.03 ± 0.10 over the GR prediction.

The existence of a 2.24" deflection angle by Freundlich, Von Kluber and Von Brunn therefore implies a requirement for the modification of the general theory of relativity. Science has evolved in this simpler manner of modifications although there are some who cling to the old thoughts of "The earth is the center of the universe and Einstein is always right". I am not proving anyone wrong but I want you to believe that the general relativity theory that was put forward by Einstein is not the only 'there is' excellent description of the universe, there are other ways far better than GR as it was with the Newtonian

Gravitational force replacement with a curvature of space time.

The introduction of a number **4** in Einstein deflection angle of light has no basis as to how it came along. The fact that his formula resembles the Newton formula actually shows that Einstein borrowed ideas from Newton analysis. He Einstein also failed to eliminate the mass of a photon from his equations. Even today no one knows how to deduce the deflection angle without taking into account the photon mass because we know the photon is massless.

Ladies and gentlemen, let me present to you another approach that will lead us to the Einstein deflection angle without assuming that the photon has mass or kinetic energy.

Quantum Gravity Approach

Let the potential energy of the Photon according to Einstein –Planck relation be,

$$E = \frac{hc}{\lambda}$$

Where λ is the wavelength

Since ligth appears curved at a small part of the Sun's surface, then the circumference according to deBrogile is quantized in units, C=πR=λ (In case light orbits the Sun, then C=2πR). Then the energy of the photon will be given by

$$E_r = \frac{2\hbar c}{R}$$

According to relativistic quantum mechanics, a photon of momentum P, has a kinetic- energy given by, where M is the Sun's mass

$$E_B - \frac{P^2}{2M}$$

According to quantum mechanics in curved space time, space is divided into small chuncks of matter (atoms of space) with a length close to the Planck length l_p, therefore the momentum of a photon passing through these atoms of space will be given by,

$$P = \frac{\hbar}{l_p}$$

This momentum is proof that the photon has no mass and what we percieve as the heaviness of the photon is actually the discrete nature of space.

Due to the discrete nature of space, there is a delay in time at which the photon will reach our telescopes from the distant star. In other words the speed of light doesn't change but there is a huge difference from the calculated time and the observed time of reach of light from the distant star. Then the energy carried by a photon through the discrete space is given as

$$E_B = \frac{\hbar c^3}{2GM}$$

This then brings us to the deflection angle which is the ratio of the photon potential energy to the kinetic energy,

$$\theta = \frac{E_r}{E_B}$$

$$\theta = \frac{4GM}{c^2 R}$$

The Extra Dimension Approach

In higher dimensions or extra dimension problems we get a different picture of what general relativity really is. We assume that light behaves differently in various dimensions and the observations of light from a distant star will vary according to the flux in the extra dimensions because it is this loss of flux to the extra dimensions which makes gravity weak yet it is strong. Therefore what determines our observations is the flux in the extra dimensions as expressed in our model below,

Let the deflection angle of light at the sun's limb be given by,

$$\theta = \frac{1}{\alpha^{n/2}}\left(\frac{R_s}{R}\right) \qquad (13)$$

Where, $R_s = \frac{2GM}{c^2}$ is the Schwarzschild radius of a gravitating body, α is the size of the extra dimension and $\alpha^{n/2}$ is the flux in the extra dimension. In what follows, we use the above equation by substituting in the values of $\alpha^{n/2}$ to get the values of the three deflection angles whose sample mean gives the Einstein deflection value. This analysis will help us recover new theories based on the flux in the extra dimension.

Let us start with the Newton's theory of gravitation. To recover the Newtonian deflection angle at the suns limb, we set $\alpha^{n/2} = 1$. This then gives the Newtonian value as,

$$\theta_N = \frac{R_s}{R_\odot} = 0.875 \text{arcsec}$$

The Freundlich deflection angle might have taken a different twist than with Eddington 1.75arcsec result, which we are yet to find out. Taking, $\alpha^{n/2} = 0.0233$, we deduce the deflection angle given by,

$$\theta_F = \frac{2.56 R_s}{R_\odot} = 2.24 \text{arcsec}$$

Lastly when $\alpha^{n/2} = 0.0290$ we get the following deflection angle,

$$\theta_Q = \frac{2.426 R_s}{R_\odot} = 2.12 \text{arcsec}$$

Our first result from the above calculations is that; the sample mean of the deflection angles from the three observations gives the exact deflection angle that was calculated and observed by Eddington in General relativity as,

$$\frac{\sum_{n=1}^{4} \theta_n}{3} = \frac{0.875 + 2.24 + 2.12}{3} = 1.75 \text{arcsec}$$

The fact that the mean of the three observations for the deflection of light given above reproduces the GR value of 1.75arcsec sums up what exactly general relativity really is. In simple terms GR is the sample mean of three observations taken from different location on the earth's surface where the flux in the extra dimension makes the strength of gravity slightly different in those positions where light bends.

The model given above is proof that the curvature of space assumption given by General Relatitiy was just a mathematical artifact and not a real entity. The observed deflection angles are greatly determined by the flux in the extra dimensions.

9. A New Approach to Quantum Theory

The Wave and Particle Analogy

Basing our study on the electric currents generated whenever there is a changing magnetic field (B) and a changing electric field (E) in the electromagnetic wave we can construct a complete theory for the electromagnetic radiations. The theory is created using the symmetry between a long wire placed in the electromagnetic fields which induce vibrating electrons that carry current in the wire and the electromagnetic wave which constitute changing electric and magnetic fields that create vibrating photons in the wave.

Therefore a wire is to a wave what a vibrating electron is to a vibrating photon in the wire and a wave respectively. The aim of the paper is to give a clear description of the theory of electromagnetic radiations (light).

The goal of the book on the other hand is to show that the wave-particle descriptions of reality can be applied to any physical situation simultaneously.

The objective of the book is to show that the Photoelectric Effect and the Compton Effect can both be explained by the wave model and the particle model at the same time.

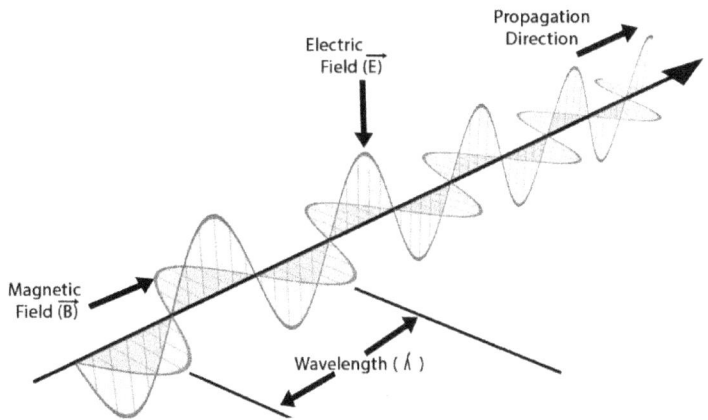

Consider a long wire connected to an ammeter and strong electric and magnetic fields produced in a vacuum. Let us assume that whenever a wire is brought in vicinity of a changing electric field, electrons of mass (m) are set into motion in the wire and then an ammeter deflects, recording a current ($\dot{i}_E$). The current in the wire due to a changing electric field should be given by

$$\dot{i}_E = \frac{j\varepsilon_o}{2\pi m} E \qquad (1)$$

Where ($\mathcal{E}_o$) is the permittivity of free space and (j) is the constant of action in SI units Js. therefore the current is quantized and depends on both the electric field and the mass of an electron.

When the wire is brought into the magnetic field, vibrating electrons at a frequency of oscillation (f) are set in motion at a speed (v) through the wire generating a current given by

$$i_B = \frac{v}{2\pi \mu_o f} B \qquad (2)$$

Where (μ_o) is the permeability of free space.

Assuming that the ammeter records different values of (i_E) and (i_B), what will be the change in the current values recorded at the ammeter? Subtracting equation (1) from equation (2) we have

$$\Delta I = (i_E - i_B) = \left(\frac{j\varepsilon_o}{2\pi m} E - \frac{v}{2\pi \mu_o f} B\right) \qquad (3)$$

This is the change in the currents due to changing magnetic and electric fields. Assuming that there is no change in the current, meaning that the current values for i_E are equal to those of i_B (i.e $\Delta I = 0$). This will imply that the magnetic field strength was equal to the electric field strength at one point in both experiments. In terms of electromagnetic radiations in the vacuum, assuming that a wire carrying current is replaced by a wave and electrons are replaced by photons. The wire replaced by a wave is made up of vibrating electric and magnetic fields at a given frequency making an electromagnetic wave. The electrons replaced with photons will represent the particle properties of the electromagnetic wave (light) with associated mass and speed (v).

The symmetry here is between the long wire and the wave, the electrons and the Photons. The electric and magnetic fields brought in vicinity of the wire and the number of oscillations per second of the electron in the wire is what leads to an electromagnetic wave. The electrons with a given mass and moving at a given speed is what constitute a photon. Then at $\Delta I = 0$, we have on arranging,

$$\frac{jf}{mv} = \frac{1}{2\pi \mu_0 \varepsilon_0} \frac{B}{E} \qquad (4)$$

This means that at $\Delta I = 0$, either a changing magnetic field or a changing electric field produces a current. Then it should be true that a changing magnetic field produces an electric field just as a changing electric field produces a magnetic field. This process in the electromagnetic wave continues indefinitely. The electromagnetic wave will move at a constant speed (c), since for electromagnetic waves, $\frac{E}{B} = c$, and for a photon $\frac{jf}{mv} = c$ where j=6.63× 10^{-34} Js (also called the Planck constant after Max Planck) and mv is the photon momentum. Implying that the photon energy is related to the frequency of the electromagnetic wave by (jf). Then the electromagnetic wave will move at a constant speed given as, since by symmetry $\frac{E}{B} = \frac{jf}{mv} = c$

$$c = \frac{1}{\sqrt{\varepsilon_o \mu_o}} = 2.99792458 \times 10^8 \frac{m}{s}$$

Where $\varepsilon_o = 8.85418782 \times 10^{-12} \frac{c^2}{Nm^2}$ and $\mu_o = 1.26 \times 10^{-6} \frac{Ns^2}{c^2}$

We have therefore deduced based on the symmetry between a current (electron) carrying wire in the electromagnetic field and the photons in electromagnetic waves that an electromagnetic wave moves at a constant speed of light. It is also true from the deductions that light is indeed made up of particles of light called photons and vibrating electric and magnetic fields. The deduction would not be possible if the wave and particle descriptions of the situations had not been applied simultaneously (into what is called "the wave-particle duality).

The Photoelectric Effect

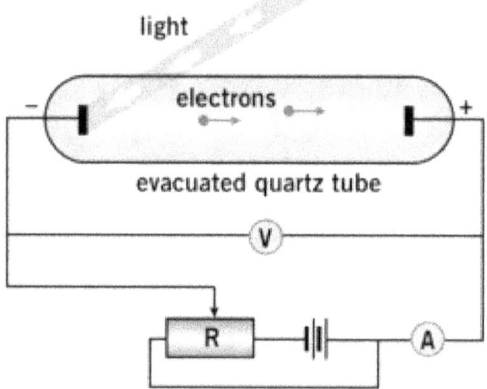

Unexpectedly enough the photoelectric effect can also be explained by Equation (3), on arranging

$$\frac{2\pi m f}{\varepsilon_0 E}\Delta I = jf - \frac{mv}{2\pi \mu_0 \varepsilon_0}\frac{B}{E}$$

Then the total energy of the particle of light (Photon) is then given by

$$jf = \frac{2\pi mf}{\varepsilon_0 E}\Delta I + \frac{mv}{2\pi \mu_0 \varepsilon_0}\frac{B}{E} \quad (5)$$

It is therefore true that the photoelectric effect can be explained when both the particle and wave models of reality are applied in the experiment at the same time (simultaneously). The work function from Einstein's photoelectric equation (A. Einstein, 1905) will here be replaced by $\frac{2\pi mf}{\varepsilon_0 E}\Delta I$ while the kinetic energy of the electrons at the surface of the metal will be given by $\frac{mv}{2\pi \mu_0 \varepsilon_0}\frac{B}{E}$. Equation (5) reduces to Einstein's Photoelectric effect when, the speed of the electron is $v = \frac{1}{\pi \mu_0 \varepsilon_0}\frac{B}{E}$ and the change in current for a complete circuit is $\Delta I = \frac{j\varepsilon_0 E}{2\pi m}$.

The Compton Effect

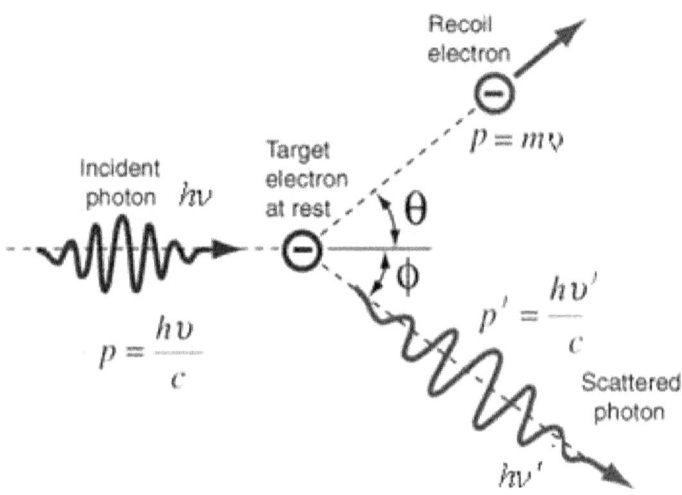

The validity of the Compton Effect can also be deduced from Equation (3). The current can be taken as the product of the frequency (f) of radiations and the charge (q) on the particle.

Then the current due to the electric field is $i_E = qf_1$ and that due to the magnetic field is $i_B = qf_2$. In the case of the Compton Effect, q is the charge on the free electron while f_1 and f_2 are the frequencies of the incoming photon and outgoing photon after collision with the free electron respectively. Then equation (3) can be written as

$$f_1 - f_2 = \frac{1}{q}\left(\frac{js_0}{2\pi m}E - \frac{v}{2\pi \mu_0 f}B\right) \qquad (6)$$

Since photons move with the speed of light(c) then their frequencies is related to their speed and wavelength by $f = \frac{c}{\lambda}$, then we have

$$\frac{1}{\lambda_1} - \frac{1}{\lambda_2} = \frac{1}{qc}\left(\frac{js_0}{2\pi m}E - \frac{v}{2\pi \mu_0 f}B\right)$$

On arranging to include the charge density of the free electron for electric field lines in an area of $\frac{\lambda_1 \lambda_2}{2\pi}$, we obtain

$$\frac{2\pi q}{\lambda_1 \lambda_2}(\lambda_2 - \lambda_1) = \frac{j}{mc}\left(\varepsilon_0 E - \frac{mv}{\mu_0 jf}B\right)$$

Where (mc) is the momentum of an electron treated relativistic ally, on letting the charge density $= \frac{2\pi q}{\lambda_1 \lambda_2} = \varepsilon_0 E$, we deduce the change in the wave length of the incoming photon and outgoing photon after collision with the free electron as

$$\Delta\lambda = (\lambda_2 - \lambda_1) = \frac{j}{mc}\left(1 - \frac{mv}{p\mu_0 jf}B\right)$$

Since $p = \varepsilon_0 E$, we then have

$$\Delta\lambda = (\lambda_2 - \lambda_1) = \frac{j}{mc}\left(1 - \frac{\frac{mvB}{\varepsilon_0\mu_0 E}}{jf}\right)$$

Since jf is the energy carried by the photon, and then also $\frac{mvB}{\varepsilon_0\mu_0 E}$ is the energy carried by the free electron. Treating the electron relativistically such that for electromagnetic waves moving at a speed (v) relative to the electron moving at a speed of light $c = \frac{1}{\sqrt{\varepsilon_0\mu_0}}$, the electric field in the wave will be related to the magnetic field by $Bv = E.$ then the energy carried by an electron can be given by mc^2. Then the angle at which the photon is scattered after collision with the free electron will be given by

$$\theta = \cos^{-1}\left(\frac{mvB}{\varepsilon_0\mu_0 E}\right)\bigg/jf \qquad (7)$$

Where mv is the momentum of the photon in the electromagnetic wave consisting of a changing electric field E and magnetic field B both moving at a constant speed of light $c = \frac{1}{\sqrt{\varepsilon_o \mu_o}}$. Treating the electron relativistically we have

$$\theta = \cos^{-1}\frac{mc^2}{jf}$$

When the energy carried by the photon is equal to the energy possessed by the electron then $\theta = 0$, meaning that there is or there is no scattering and whatsoever there is no increase in photon wavelength hence $\Delta\lambda = 0$.

Bohr's Atomic Theory

A complete theory of light can't fail to explain the structure of an atom. I therefore take a complete discussion of what goes on inside an atom only with the help of Bohr's energy levels which he derived using classical mechanics and quantum theory. Let $\Delta f = f_1 - f_2$ be an increase in the frequency of the electromagnetic radiations emitted from an atom. Then squaring both sides of equation (6) and arranging will give

$$4\pi^2 \Delta f^2 = \frac{1}{m^2 q^2}\left(j\varepsilon_o E - \frac{mv}{\mu_o f}B\right)^2$$

$$4\pi^2 m^2 q^2 \Delta f^2 = j^2 \varepsilon_o^2 E^2 - 2\frac{j\varepsilon_o EBmv}{\mu_o f} + \frac{B^2 m^2 v^2}{\mu_o^2 f^2}$$

Dividing through by $64\pi^4 j^2 \varepsilon_o^2$ and multiplying through by q^2 gives the energy of the atom as on arranging

$$\frac{mq^4}{16\pi^2 \pi^4 j^2 s_o^2} = \frac{1}{64\pi^4 m\Delta f^2}\left((Eq)^2 - 2\frac{m(Eq)(Bqv)}{\mu_o \varepsilon_o (jf)} + \frac{(Bqv)^2 m^2}{\mu_o^2 \varepsilon_o^2 (jf)^2}\right)$$

The energy of the n-th level is since the reduced Planck constant is

$$n\hbar = \frac{nj}{2\pi}$$

$$\frac{mq^4}{32\pi^2 n^4 n^2 \hbar^2 \varepsilon_0^2} = \frac{1}{32\pi^2 n^2 m\Delta f^2}\left((Eq)^2 - 2\frac{m(Eq)(Bqv)}{\mu_0\varepsilon_0(jf)} + \frac{(Bqv)^2 m^2}{\mu_0^2\varepsilon_0^2(jf)^2}\right)$$

The expression on the left hand side of the equation is the quantized energy of an atom (Niels Bohr, 1913) while the right hand side of the equation represents the energy of the atom in terms of the forces associated with it. In the equation we let $H_e = Eq$ be the electric force for a particle moving in the electric field and $H_b = Bqv$, the magnetic force on a particle with charge q moving in the magnetic field. Since the speed of light is $= \frac{1}{\sqrt{\varepsilon_0\mu_0}}$, then the quantized energy can be given as

$$W_n = \frac{1}{32\pi^2 n^2 m\Delta f^2}\left(H_e^2 - 2\frac{H_e H_b mc^2}{jf} + \frac{H_b^2(mc^2)^2}{(jf)^2}\right)$$

Then on arranging we obtain

$$W_n = \frac{1}{32\pi^2 n^2 m\Delta f^2}\left(H_g - \frac{mc^2}{jf}H_b\right)^2 \qquad (8)$$

When the energy of an electron moving at a speed of light in atom is equal to the energy of the emitted photon, then

$$W_n = \frac{1}{32\pi^2 n^2 m\Delta f^2}(H_g - H_b)^2 = \frac{1}{32\pi^2 mn^2}\left(\frac{\Delta H}{\Delta f}\right)^2 \qquad (9)$$

Where $\Delta H = H_g - H_b$ is the difference or change between the electric force and the magnetic force in an atom, when the two forces balance (i.e. $H_g = H_b$), then $W_n = 0$ meaning that the total energy of an atom will cease to exist.

Therefore the total energy of an atom increases with the square of the change in the electric and magnetic forces which govern an electron but falls off as the square of the change in the frequency of the radiation emitted by it.

From equation (8) the ratio of the energy of an electron to that of the photon $\frac{mc^2}{jf}$, is the limit at which if the energies are not equal you will not get a change in the electric and magnetic forces. Treating the ratio as a number $\tau = \dfrac{mc^2}{jf}$, we get from equation (8)

$$W_n = \frac{1}{32\pi^2 mn^2}\left(\frac{H_g - \tau H_b}{f_1 - f_2}\right)^2 \qquad (10)$$

When $\tau = 0$, it means that the relativistic energy (mc^2) of an electron in an atom is zero, and that the total energy of an atom only increases with the electric force on the electron. The relationship (equation 10) is a complete expression for the laws according to which, by the theory here advanced, the structure of an atom should be viewed.

In conclusion, a complete theory of light is only possible if both the wave and particle descriptions of reality are applied to the physical situation at the same time. In discussing Young's double slit experiment for example we should be able with the formulas given above to treat the electromagnetic radiations on both a wave and particle model.

10. Vacuum Pressure Gravity as a Possible Alternative to Newton's law of Gravity

In Einstein's theory of gravitation one postulates that gravity is not a force, but it is really an effect caused by the curvature of space and time.

$$\Lambda = \frac{8\pi G}{c^2}\rho \qquad (1)$$

Where, Λ is the cosmological constant (curvature of spacetime) and ρ is the pressure –energy density (matter energy content of the universe)

According to equation (1), matter curves space-time in its vicinity and this curvature in return affects how matter moves.

Here we consider the hypothesis that, what we call space-time (a mathematical description telling nothing about the physical mechanism starting the motion) is replaced by a physical vacuum fluctuating and generating particle pairs (representing zero point energy of vacuum) that appear and disappear incredibly.

This zero point energy of physical vacuum is the dominant energy of the universe and the Casmir or vacuum pressure of this energy is the main cause of the observed accelerated expansion of the universe. In other words, virtual particles drive the accelerated expansion of the universe according to the new law given below;

$$F = \sqrt{4\pi\alpha\hbar c p} \qquad (2)$$

Where, F is the attractive force between any two points in the vacuum of space, p is the vacuum pressure-energy density, α is the coupling, $\hbar$ is the reduced Planck constant and c is the constant speed of light.

According to Equation 2 above, the gravitational force experienced by a particle in a vacuum is proportional to the square root of the vacuum pressure experienced by a particle.

This hypothesis is not only a modification of Newton's law and Einstein General Relativity to account for the observed properties of galaxies but is also a law to be applied in the study of the behavior of particles on a microscopic scale. It is also an alternative to the hypothesis of dark matter in terms of explaining why galaxies do not appear to obey the currently understood laws of physics. For example applying this force to an object of mass m in circular orbit around a point mass M for a star in the outer regions of a galaxy we find;

$$\frac{mv^2}{r} = \sqrt{4\pi\alpha\hbar cp}$$

But vacuum pressure $(P) = \frac{Force\ (F_o)}{Area\ (A)}$

According to Newton's second law $F_o = ma_o$

Where, a_o is centripetal acceleration of a star as opposed to the centripetal acceleration itself, as in Newton's second law. This force is perpendicular to the surface area $A = 4\pi r^2$ enclosed by the boundary of orbit

$$\frac{mv^2}{r} = \sqrt{\frac{\alpha\hbar cma_o}{r^2}}$$

The gravitational coupling is, $\alpha = \frac{GMm}{\hbar c}$

Squaring both sides of the equation and arranging we have;

$$v^4 = GMa_o$$

That is, the star's rotation velocity is independent of r, its distance from the centre of the galaxy-the rotation curve is flat, as required. This is the best prediction of MOND- the Tully-Fisher relation.

However when, $F = F_o$ we recover Newton's law of gravity

This means that, in the outer regions of a galaxy, Newton's law of gravity no longer applies because the gravitational force is inversely proportional to the radius (as opposed to the inverse square of the radius, as in Newton's law of gravity)

$$F = \frac{1}{r}\sqrt{\alpha \hbar c F_o}$$

$$\Rightarrow \frac{F^2}{F_o} = \frac{ma^2}{a_o} = \frac{\alpha \hbar c}{r^2} = F(casimir\ force)$$

a_o is a new fundamental constant which marks the transition between the Newtonian and deep MOND regimes.

This implies that gravity is a result of the Casimir pressure effect.

11. The Theory of Everything

Physicists have argued out that the more elegant and symmetrical the theory is, the more it is beautiful. The elegancy of any physical theory is suspected at a level to which it holds well with other theories , that is ,the capability of the theory to conform with the well known laws of nature at all levels.

In this section we examine the mechanism through which quantum mechanics becomes comparable with gravity and the scale at which this occurs. At the Planck scale all interactions (the weak interaction, strong interaction and electromagnetism) are assumed to merge into a single interaction that alone occurs at very high energies of about 1TeV. The equations that do describe this phenomenon are not yet found and therefore requires one's deep effort to capture the reality of this entire puzzle.

To capture interest in these interactions we need to know first, their strength and second the range in which they occur. The strength defines the coupling constants and the range defines the attractions, on the other hand the coupling constant determines the strength of any interaction and therefore is a number in a sense that it is a dimensionless constant. A coupling constant is a very important quantity in dynamics, for example, in the motion of a large lump of magnetized iron, the magnetic forces are more important than the gravitational forces because of the relative magnitudes of the coupling constants.

The standard model is a theory of three fundamental forces - electromagnetism, weak interactions and strong interactions; however, these three forces are not tied together Howard Georgi and Sheldon Glashow discovered that the Standard Model particles can arise from a single interaction, known as a grand unified theory. Grand unified theories predict relationships between otherwise unrelated constants of nature in the Standard Model. Gauge coupling unification is the prediction from grand unified theories for the relative strengths of the electromagnetic, weak and strong forces and this prediction was verified at LEP in 1991 for supersymmetric theories.

In particle physics, supersymmetry (often abbreviated SUSY) is a novel symmetry that relates elementary particles of one spin to another particle that differs by half a unit of spin and are known as superpartners. Since the particles of the Standard Model do not have this property, supersymmetry must be a broken symmetry allowing the 'sparticles to be heavy.

One of the main motivations for SUSY comes from the quadratically divergent contributions to the Higgs mass squared. The quantum mechanical interactions of the Higgs boson causes a large renormalization of the Higgs mass and unless there is an accidental cancellation, the natural size of the Higgs mass is the highest scale possible. This problem is known as the hierarchy problem Supersymmetry reduces the size of the quantum corrections by having automatic cancellations between fermionic and bosonic Higgs interactions. If supersymmetry is restored at the weak scale, then the Higgs mass is related to supersymmetry breaking which can be induced from small non-perturbative effects explaining the vastly different scales in the weak interactions and gravitational interactions. The failure of experiments to discover either supersymmetric partners or extra spatial dimensions, as of 2006 has encouraged loop quantum gravity researchers.

(a) The determination of the strength of the forces

We assume a model that explains everything on the length scales, the best scale so far we are familiar with is the Planck length scale, however in this model we don't associate ourselves in knowing this scale and therefore develop new scales that alone are combined together to lead to some observable phenomenon describing the forces involved in the interactions. The equation describing the model is developed and given by;

$$(v^2/c^2 + n^2\beta_{Qo}) = 8\pi\beta_{gEo} \qquad (1)$$

Where β_{Qo} is a length ratio given by l_Q/l_o, in this case $l_Q = \hbar c/W$, $\hbar$ is Dirac constant, c is the speed of light and W is the energy. Also $\beta_{gEo} = l_{gE}/l_o$ where $l_{gE} = 8\pi G M_{gE}^2/W$, G is the universal gravitational constant, M_{gE} is the mass of a particle in the combined fields given by P_{gE}/c where P_{gE}

$=Gm^2ke^2/R^2c^2$, is the momentum for an elementary particle of mass m and an elementary charge e, k is the coulomb constant and R is the distance between any two particles. The equation here addresses the problems in form of length scales simply because it is at these scales that quantum mechanics seem to be comparable to gravity. The momentum P_{gE} is a momentum of a particle experiencing the strength of the electromagnetic fields and gravity. The strength is determined by a very small coupling constant as we shall see later. The smaller the distance between elementary particles, the higher the momentum and vice versa is true.

The exchange of photons between an electron and a proton in an atom is explained by Quantum Electrodynamics (QED), with a coupling constant determining the strength of the electromagnetic force. The equation of the interaction responsible for QED on the length scale, which is the Compton length, is given by the equation

$$\sum \psi^2_{fi} t_i = 2\pi \beta_{RCE} \qquad (2)$$

The expression $\sum \psi^2_{fi} t_i$ is the force changer where $\psi_{fi} = f_i R^2/ke^2$ and $t = \{f_i^2 ke^2/R^2\}/F_n^3$,

β_{RCE} remains a constant given by l_c / l_{RE} (l_c is the Compton length $\hbar/mc$ and $l_{RE}=ke^2/mR^2c^2$).On multiplying both sides of Eqn1 by a quantity $\sum\psi^2_{fi}t_i$ we obtain,

$$(v^2/c^2 + n^2\beta_{Qo}) \sum\psi^2_{fi}t_i = 8\pi\beta_{gEo}\sum\psi^2_{fi}t_i$$

We then examine the condition for which β_{Qo} will be a maximum and minimum. It is found out from relativity that β_{Qo} is maximum when the lorentz factor $\gamma = (1-v^2/c^2)^{-1/2}$ is very small that is,

$\gamma = 1/n\sqrt{\beta_{Qo}}$ or when the velocity $v = c \sqrt{(\sum\psi^2_{fi}t_i - n^2\beta_{QO})}$

We hence obtain a general interaction equation as,

$$\sum f_y^3\psi^2_{fi}t_i = 2\pi\beta_{RCE}\sum F_n^3/\xi\beta_{gEo}, \quad n = 1,2,3 \qquad (3)$$

The following conditions are then taken into account

1) For $l_o = l_x = mc^2/F_p$, $\beta_{gEo} = \beta_{gEx} = l_{gE}/l_x$.

2) For $l_o = l_c$, $\beta_{gEo} = \beta_{gEQ} = l_{gE}/l_c$, and

3) For $l_o = l_s = Gm/c^2$, $\beta_{gEo} = \beta_{gEs} = l_{gE}/l_s$, which gives

$$\sum f_y^3 \psi^2_{fi} t_i = F_1^3 + F_2^3 + F_3^3 = 2\pi \beta_{RCE} (F_p^3/8\pi\beta_{gEx} + F_p^3/256\pi^3\beta_{gEQ} + 2F_B^3/\pi\beta_{gEo}) \quad (4)$$

Where $F_p = c^4/G$ is the Planck unit force and $F_B^3 = m^2c^3/\hbar$ is the force required for strong and weak interactions to take place. Again setting a condition,

For for $l_o = l_z = ke^2/mc^2$, $\beta_{gEo} = \beta_{gEz} = l_{gE}/l_z$.

$$\sum f_y^3 \psi^2_{fi} t_i = F_4^3 = 2\pi\beta_{RCE} (F_z^3/32\pi^3\beta_{gEx})$$

Where $F_z^3 = m^2c^4/ke^2$,

Also for $l_o = l_N = \hbar^2 m^3 G^2/k^3 e^6$, $\beta_{gEo} = \beta_{gEN} = l_{gE}/l_N$, we obtain,

$$\sum f_y^3 \psi^2{}_{fi} t_i = F_5^3 = 2\pi \beta_{RCE} (F_z^3/2\pi^3 \beta_{gEN}) \quad (5)$$

Measuring the value of the strong, weak and electromagnetic coupling constants gives us away through which we can determine supersymmetric levels. From supersymmetry and grand unification of elementary particles the couplings agree to 1%. The relationships of the sum of the cubes of the forces to each individual cube of the force, and that of the sum of the square of masses with each known mass squared casts much information about the masses and couplings of the supersymmetric particles as shown below, when Eqn4 is divided through respectively by the cubes of the forces F_1^3, F_2^3 and F_3^3 the following equations are obtained,

$$\sum F_n^3/F_1^3 = 1 + 16\alpha_g^3 + 1/32\pi^2 \alpha_g \quad (6)$$

$$\sum F_n^3/F_2^3 = 1 + 32\pi^2 \alpha_g + 512\pi^2 \alpha_g^4 \quad (7)$$

$$\sum F_n^3/F_3^3 = 1 + 1/6\alpha_g^3 + 1/512\pi^2 \alpha_g^4 \quad (8)$$

$$\sum F_n^3/F_4^3 = 1 + \beta^2(4\pi^{2+}1/8\alpha_g) + 64\pi^2\alpha_s^2\alpha_g \quad (9)$$

Where $\beta = ke^2/Gm^2$ is the ratio of the fine structure constant α_s to the gravitational coupling constant α_g, given respectively as $\alpha_g = Gm^2/\hbar c$ and $\alpha_s = ke^2/\hbar c$.

Now equating $F_4 = F_5$, $F_5 = F_3$, $F_5 = F_1$ we obtain; m_1, m_2, m_3 and m_4 respectively, Adding the squares of the masses we obtain,

$$\sum m_n^2 = m_1^2 + m_2^2 + m_3^2 + m_4^2 \quad (10)$$

Which gives the sum per unit mass as,

$$\sum m_n^2/m_1^2 = 1 + 16\pi^2\alpha_s^4 + (8\pi/\alpha_s)^{\frac{1}{2}} + 4/(128\,\alpha_s^4)^{1/5} \quad (11)$$

$$\sum m_n^2/m_2^2 = 1 + 1/16\pi^2\alpha_s^4 + (1/8\pi\,\alpha_s^9)^{\frac{1}{2}} + (1/4\pi^2)(128\,\alpha_s^{24}) \quad (12)$$

The equations generated so far give a basis for the nature and type of supersymmetry exhibited by a particle experiencing forces at both the Planck and grand unified scales. It is thus shown here that the electromagnetic coupling constant is a result of mathematically summing the squares of the masses generated and then dividing through by the square of the mass in the summation while the gravitational coupling constant is the result of summing the cubes of the forces and then dividing through by the cube of the force in the sum. This idea at its best is taken to be the basis for symmetric theories as we shall see in the results obtained.

(b)Results

(i)The unification of coupling calculations

At equal forces that is $F_1 = F_2 = F_3 = F_p$ the mass $M_p = (\hbar c /8\pi G)^{1/2} = 2.1765 \times 10^{-8}$ kg, is obtained which is the Planck mass for which the Schwarzschild radius is equal to the Compton length divided by π. When Eq4 is divided through by F_1^6 and F_2^6 we obtain equations of the form;

$$\sum F_n^3/F_1^6 = \Omega/F_p^3 \quad (13)$$

$$\sum F_n^3/F_2^6 = \epsilon/F_p^3 \quad (14)$$

Where, $\Omega = 4m^2/m_p^2 + 1/\pi + m^8/32\pi^2 m_p^8$ and $\epsilon = m^6/8\pi m_p^6 + 16\pi m^4/m_p^4 + 2m^{12}/\pi m_p^{12}$

The mass relations equations obtained above indicate the scale at which gravity may be strong and weak. Obtaining these results on the Planck force and mass scale is evidence for the existence of the theory of quantum gravity. The values Ω and ϵ represent a series equation defined by increasing powers in the mass ratio (**m/m$_p$**). The mass **m** is assigned to any particle and the mass **m$_p$** is assigned to the Planck scale defining quantum gravity.

The unit of energy is **M$_P$c^2**; the unit of electric charge is **√hc/k**, where k is coulomb constant and so forth. On the other hand, one cannot form a pure number from these three physical constants. Thus one might hope that in a physical theory where ℏ, c, and G were all profoundly incorporated, all physical quantities could be expressed in natural units as pure numbers. Within its domain, this paper has achieved it for example, imagining that there were just two quark species with vanishing masses. Then from the two integers 3 (colors) and 2 (flavors), ℏ, and c (without mass parameters), the spectrum of hadrons with mass ratios and other properties close to those observed in reality, emerges by through calculation (Ω and ϵ) as indicated from Eqn13 and Eq14 shown above. The overall unit of mass is indeterminate, but this ambiguity has no significance within the theory itself. The results obtained show an ideal Planckian theory that alone does not contain any pure numbers as parameters. Thus, for example, the value $m_e/m_p = 10^{-22}$ of the electron mass in Planck units is obtained from a dynamical calculation. This ideal might be overly ambitious, yet it seems reasonable to hope that significant constraints among physical observables will emerge from the inner requirements of a quantum theory which consistently incorporates gravity. The model therefore provides; first, the unification of couplings calculation. second, it points to a symmetry breaking scale remarkably close to the Planck scale (though apparently smaller by 10^{-2} to 10^{-3}), so there are pure numbers with much more 'reasonable' values than 10^{-22} to shoot for. Third, it shows quite concretely how very large scale factors can be controlled by modest ratios of coupling strength, due to the logarithmic nature of the running of couplings (so that 10^{-22} may not be so 'unreasonable' after all).

While the above result is based on the study of the strength of the gravitational force, we now look for ways in which we can examine the strength of the electromagnetic force depending on the mass. This is done by dividing the sum of the squares of the masses (Eqn10) by the fourth power of the individual masses hence,

$$\sum m_n^2/m_2^4 = \omega/m_G^2 \quad (15)$$

$$\sum m_n^2/m_E^4 = \lambda/m_p^2. \quad (16)$$

Where $\omega = 1/\ 16\pi^4\alpha_s^9 + 1\ /4\pi^2\alpha_s^5 + 1/\ (512\pi^8\ \alpha_s^{19})^{1/2}$,
$\lambda = 128\pi^3\alpha_s^6 + 8192\pi^5\alpha_s^{10} + 128\sqrt{\pi^3}\alpha_s^{11} + 128(\pi^{15}\alpha_s^{26})^{1/5}$

$m_E = (1/\ 8\pi\ Ke^2)(\hbar^3 c^3/G)^{1/2}$ is the mass obtained when $F_4^3 = F_3^3$, and $M_G = (Ke^2/G)^{1/2}$ is the mass obtained when the electromagnetic force is equal to the gravitational force.

It can now be theorized that the strength of the electromagnetic force is determined by Eqn15 and 16 at which a series power equation in the fine structure constant defined by ω and λ is a constant.

(ii) The length scales at which the masses predicted by the standard model survive

The mass of the W and Z bosons (M_W, M_Z), Higgs particle (M_H) and the mass scale at the grand unification (M_{GUT}) are generated. We multiply a coupling constant μ with the force F_3^3, of which we equate to F_4^3 that is;

$$\mu F_3^3 = F_4^3$$

From which

$$\mu = R_B^2 / R_o^2$$

Where R_o is the length scale determined experimentally and $R_B = (8\pi G k e^2/c^4)^{1/2} = 6.9101 \times 10^{-36}$ m, which is greater than the Planck length.

So the equation that produces the different masses at R_o will be given by the square of the mass as,

$$M^2 = m_p^2 / 8\pi\mu \alpha_s^2$$

Where $\alpha_s = 1/137$, is the electromagnetic coupling constant.

To obtain the masses, we need to find the length R_o, theoretically we develop the lengths given by; 1.03741×10^{-39} m, 8.3182×10^{-54} m, 9.4334×10^{-54} m, and 1.2345×10^{-53} m.

Following the given lengths we respectively obtain the masses;

$M_{GUT}=10^{16}$ GeV, $M_W = 80.18$ GeV, $M_Z = 90.82$ GeV, and $M_H = 119$ GeV respectively.

But at $R_B = R_o$, the mass $M_B = 6.661 \times 10^{19}$ GeV is obtained. And at $R_o = 2.529 \times 10^{-37}$ m, the Planck mass is obtained (that is $M = m_p$). Therefore it is found out that the W and Z boson particles survive in length of 10^{-54} m. The Higgs particle survives to a length greater than that of the W boson $\geq 10^{-53}$ m. And finally particles at the grand unified scale will survive at 10^{-39} m.

(iii) The big bang acceleration and proton decay

For proton decay the intensity **P** is used such that at Schwarzschild radius R and Planck mass scale m_p the life time of the proton as explained by SUSY is seen to agree so well with the

$$T(time) = \alpha^2 \, m^5_p \, R / 4096 \, \pi^3 \, m_k^4 \, \hbar$$

Such that at $m_k = 7.96\times10^{-29}$ GeV, $T = 10^{35}$ yrs.

We have obtained the lifetime of protons and the mass of a particle produced during the decay process. The mass of the particle obtained is very small and can therefore be taken to be a neutrino.

The force F_3 can be expressed in the form,

$$F_3 = a_3(m_3^5 / 16\pi^2 m_p^2)^{1/3}$$

Where a_3 is the acceleration, this acceleration at a Planck scale will be given by

$$a_3 = (c^{11} / \hbar G^2 m)^{1/3} = 2.4772\times10^{52} \text{m/s}^2$$

This is quite a very large acceleration and therefore defined as the acceleration of particles during the early formation of the universe.

The results obtained describe super symmetry which is a theory required for the unification of everything we know about the physical world into a theory of everything. Significantly a larger enterprise of the theory is to produce a theory of quantum gravity which is required for the unification of general relativity with the standard model, which explains the other three basic forces in physics (electromagnetism, the strong interaction, and the weak interaction), and provides a palette of fundamental particles upon which all four forces act. Theoretically the results obtained (Eqn11 and Eqn12) show a huge correction to the particles' masses, which without fine-tuning will make them much larger than they are in nature. The problem of the unification of the weak interactions, the strong interactions and electromagnetism is solved mathematically, through the comparisons of the cube of the forces in a ratio that generates the gravitational coupling constant power equation.

The Planck mass is the mass of a black hole whose Schwarzschild radius multiplied by π equals its Compton wavelength. The radius of such a black hole is roughly the Planck length, which is believed to be the length scale at which both general relativity and quantum mechanics simultaneously become important. In accordance with the results obtained it is seen that the Planck mass is the mass at which the four forces (F_1, F_2, F_3 and F_p) are equal, the forces are then taken to be related to the origin of the universe simply because at those high energies that formed the dense soup of the universe the forces were equal and the masses probing the Planck mass scale that is black holes were produced, hence those four forces a significant in that they play a crucial role in the formation of black holes. The intensity P on the other hand explains a phenomenon that occurs at the cosmic scale, for example it explains the nature of Black holes and the age of the universe. The acceleration obtained is so large that it is the acceleration that the universe had at the instant after the big bang. Obtaining this acceleration is the possibility of studying the rate of expansion of the universe at large, the accelerating universe is therefore the observation that the universe appears to be expanding at an accelerated rate.

At the Planck scale the descriptions of subatomic particle interactions in terms of quantum field theory breaks down. Also at the same scale, the strength of gravity is expected to become comparable to the other forces, mathematically all the fundamental forces are unified at that scale. The results obtained explain both the weak and strong interactions that at a length between 10^{-37}m and 10^{-35} the Planck scale is attained also at lengths 10^{-39}m , the grand unified scale becomes relevant , but for lengths 10^{-53}m and 10^{-54}m, the standard model holds on well. We have therefore attained a unification that increases from about 10^{-59}m (standard model) to 10^{-35}m (quantum gravity). The paper there fore gives out the relationship between elementary particle physics and astrophysics at a large scale.

Basing on the results obtained, it is now clearly justified that gravity can be integrated with quantum mechanics at the Planck scale. And therefore the success of the "standard model" which includes both the electroweak theory and quantum chromodynamics can now be regarded as successful in providing accurate descriptions of the fundamental particles and their interactions.

12. What is Semi-Classical Gravity?

For the past thirteen years, I have been working on the most important theories of physics from scratch without employing the methods of general relativity and quantum field theories and I have come up with promising results. I have deduced the Black hole thermodynamics from first principles, I have deduced the Wiedmann Franz law from scratch, the Stefan Boltzmann power law, The result for the earliest period of time in the history of the Universe, I have related the Chandrasker theory of white dwarfs with the Bohr theory of the Hydrogen atom-the results are suprising, the rest is history. This book gives a clear account of these fields of physics.

The truth is, I hate Einstein and Hawking. I don't like them because I find it hard to use their mathematical ideas to deduce the theories I desire. It was that hard for me to classify where in the scientific community I fall, at first I thought that my ideas where into the quantum gravity field section but this was a lie. The quantum theory of gravity has not been fully settled. It was yesterday that I realized that my ideas fell into the Semi-Classical physical regime when I browsed it online;

"Semi-classical physics refers to a theory in which one part of a system is described quantum-mechanically whereas the other is treated classically" In general, it incorporates a development in powers of Planck's constant, resulting in the classical physics of powers 0, and the first nontrivial approximation into the powers of -1. (Wikipedia)

I am sorry, Semi-classical physics hasn't gained much interest, there are too many criticism about its meaning, researches into the field have been discouraged, few physicists have written about it and it is that unimportant. But anyway I am an amateur to venture into a field that is irrelevant. I don't give a damn what you think.

My first insight into the field of Semi- classical physics is traced back in 2010 in my first paper I published on arXiv.org titled "A hypothetical investigation into the realm of the microscopic and macroscopic universes beyond the standard model" This paper clearly shows that I was into the field without knowing. For sure I thought I was dealing with the field of Quantum Gravity by then.

Well, if you don't understand Semi-classical physics, Amateurs do. Below I show you why I think I understand the field and you surely do. I provide many ideas which I think the entire scientific community must investigate.

(a) The meaning of semi-classical physics to an amateur

Assuming an experiment where the classical electric force f_e is balanced over the classical gravitational force f_g to determine their strength, the result will show that, the ratio of the two forces will follow a power law in powers of n of the gravitational coupling constant as,

$$\frac{f_\varepsilon}{f_g} = \alpha_g{}^n$$

The left hand side of the equation represents the classical part of the system while the right hand side represents the quantum mechanical part of the system.

Let the classical part be described by two constants;

G-The Universal gravitational constant

c- The constant speed of light

Into (G, c)

Let the Quantum mechanical part be described by two constants,

$\hbar$- The reduced Planck constant

c-The constant speed of light

Into ($\hbar$, c)

Then from the above assumption Semi-classical physics will reduce results combining the constants above into (G, c, $\hbar$)

From the above formula we can deduce the time and length units of measure formulas to help us understand the field better,

$$\text{Time } t_n = \frac{Gm}{c^3} \alpha_g^{-n}$$

$$\text{Length } l_n = \frac{Gm}{c^2} \alpha_g^{-n}$$

Where m denotes the mass of a particle or body and $\alpha_g = \frac{Gm^2}{\hbar c}$ is the gravitational coupling constant. You can also assume interactions involving the electromagnetic coupling constant. The different fields of physics resulting from the above classification for different powers of n from 0, 1, 2 and -1/2 are given below,

For n=0

Classical General Relativity

$$t_0 = \frac{Gm}{c^3}$$

$$l_0 = \frac{Gm}{c^2}$$

For n=1

Quantum mechanics

$$t_1 = \frac{\hbar}{mc^2}$$

$$l_1 = \frac{\hbar}{mc}$$

For n=2

Semi-classical gravity

$$t_2 = \frac{\hbar^2}{Gcm^3}$$

$$l_2 = \frac{\hbar^2}{Gm^3}$$

For n= -1/2

Planck Units

$$t_{-1/2} = \left(\frac{G\hbar}{c^5}\right)^{1/2}$$

$$l_{-1/2} = \left(\frac{G\hbar}{c^3}\right)^{1/2}$$

The above derivation gives out a clear description of Semi-classical physics to a lay person.

One can decide to use n as a spatial dimension of space.

(b) Applications of semi-classical physics

(i) Radiation intensity of a black hole

The classical part of a system

Let the classical total force on an electron in orbit at a distance r from the nucleus of an atom be related to its electromagnetic and gravitational forces by,

$$f = \frac{F_G F_e}{F_B}$$

Where F_G is the gravitational force, F_e is the electric force and $F_B = Bev$ is the magnetic force

The angular momentum of an electron is given classically as,

$$L = \frac{Gm^2}{c} = mvr$$

The Quantum mechanical part of the system

The angular momentum is quantized as,

$$L = \frac{K_e e^2}{c} = \hbar$$

On eliminating the constant speed of light c from both the expression of the angular momentums we have

$$mvr = \frac{F_G}{F_e}\hbar$$

The ratio $\frac{F_G}{F_e}$ represents the classical part of the system while $\hbar$ represents the quantum part.

Eliminating F_G from the above expression we get the magnetic power as,

$$F_B c = \frac{2\pi r^2 \lambda m v F_e^2}{h^2}$$

But the de Brogile wave length of an electron is $\lambda = h/mv$ and the surface area of the sphere of orbit of an electron is $A = 4\pi r^2$. Then the electromagnetic Intensity is given as,

$$I = \frac{F_B c}{A} = \frac{F_e^2}{2h}$$

Thus the intensity of a wave is proportional to the square of the electric force If we let the power of the electromagnetic wave be P= F_Bc, and n be the fine structure constant α =$ke^2/\hbar c$, then the equation for the intensity of the classical electromagnetic wave comes out clearly as,

$$P = EB/\mu_o = 2\varepsilon_o E^2 c,$$

Where μo is the permeability of free space

Assuming the coupling of the forces to be,

$$\frac{f_e}{f_g} = \alpha_g^{\ n}$$

Then at n = -1, and $f_g = \frac{c^4}{8\pi G}$ we have the electric force as,

$$f_e = \frac{\hbar c^5}{8\pi G^2 m^2}$$

Then the intensity of the radiations will be given as

$$I = \frac{f_e^{\ 2}}{2h} = \frac{\hbar c^{10}}{256\pi^3 G^4 m^4}$$

This expression comes from treating the particle classically in one part and then quantum mechanically in another part. It can be clearly seen above, that we haven't used the mathematics of general relativity or quantum field theory to reach at the result.

(ii) The earliest period of time in the history of the universe

Classical part of the system

Let the acceleration due to gravity of a particle (say an electron) in the gravitational field be given as

$$g = R/\Delta t^2$$

Where is Δt the time and R is the distance of the particle from the source. If the particle radiates energy then the energy per unit time is,

$$P = c^5/G$$

Quantum mechanical part of the system

The power and time must be quantized in units of ℏ = h/2π where h is Planck constant, hence

$$P\Delta t^2 = n^2 \hbar$$

Where n= 1,2,3…….. is the principle quantum number.

But the potential energy of the electron in the various energy states is,

$$W = -ke^2/R$$

where k is the Coulomb constant and e is the elementary charge. Since Δt^2 is known from the expression for acceleration due to gravity. Then the distance R is,

$$R = n^2 Gg\hbar/c^5$$

From which the total energy is given by,

$$W = -ke^2c^5/n^2Gg\hbar$$

From the Bohr-Einstein frequency (f) condition, applied to a transition from a level with n = n_i to a level with n = n_f, The energy of a photon emitted by a hydrogen atom is given by the difference of two hydrogen energy levels

$$hf = E_i - E_f$$

Since frequency f = c/λ, where λ is the wavelength. Then we have,

$$1/\lambda = [ke^2c^4/2\pi G\hbar^2][1/g][1/n_f^2 - 1/n_i^2]$$

The equation obtained above shows some how a great significance of gravity in the quantum theory. So far it states that regardless of the levels in the transitions of an atom the acceleration due to gravity of the particles in the atom do greatly affect the nature of its spectrum.

The quantity $[ke^2c^4/2\pi G\hbar^2]$ in the formula above is the inverse of the square of time t and therefore,

$$1/t^2 = [ke^2c^4/2\pi G\hbar^2]$$

From which the time is obtained as $t = 1.58873 \times 10^{-42}$ s. This is the earliest period of time in the history of the universe.

(iii) The Weidmann Franz- Lorenz law

Treating one part of the system classically (macroscopic) and the other quantum mechanically (microscopic), we have the formula for the electric force acting on an electron in motion as

$$F = \frac{n^2}{\alpha_g} f_g$$

Where n, is the principle quantum number.

The above formula differs from the one previously given. On squaring the above equation we obtain the square of the electric field as,

$$E^2 = \frac{n^4 c^4}{G^2 e^2 m^2}\left(\frac{c^3 \hbar}{8\pi G m}\right)^2$$

From the formula for the temperature of the black hole, the function $\frac{c^3 \hbar}{8\pi G m}$ is related to temperature as kT, and then the law for thermal conductivity will be reduced as,

$$\frac{\pi^2 E^2 G^2 m^2}{3 T c^4} = \left(\frac{n^4 \pi^2}{3}\right)\left(\frac{k}{e}\right)^2 T$$

The left hand side represents the ratio of the thermal conductivity K to the electric conductivity δ. The right hand side is the Weidman –Franz law. Therefore the left side of the equation represents the macroscopic part of the system while the right hand side represents the microscopic part of the system. Then the left-hand side will be given as,

$$\frac{K}{\delta} = \frac{1}{3}\left(\frac{\pi G m}{c^2}\right)^2 \frac{E^2}{T} = \frac{\pi A\, E^2}{3\, T}$$

156

Where A is the surface area of a body $A = \pi r_s^2$ with the schwarzichild's radius r_s. This is the conductivity ratio of a black hole.

13. The Extra Dimension Problem

The big Problem: *(i) Is it true that at every spatial dimension, there exists new physics and that it is the work of Physicists to find out? What is the method or procedure through which new physics can be found? Does this require extra dimensions?*

(ii) Does nature have more than four space-time dimensions? If so, what is their size? Are dimensions a fundamental property of the universe or an emergent result of other physical laws? Can we experimentally observe evidence of higher spatial dimensions?

(iii) Can the singularities that plague the General theory of Relativity be resolved in any quantum theory of Gravity?

History tells us that if we hit upon some obstacle, even if it looks like a pure formality or just a technical complication, it should be carefully scrutinized. Nature might be telling us something, and we should find out what it is (G. t Hooft, 1997).

In physics, one of the ultimate goals is to unify the fundamental forces of nature. Today physicists have been able to unify three of the four known fundamental forces (the electromagnetic, the strong and the weak nuclear forces in a single quantum field theory-the standard model). The fourth fundamental force, gravity, on the other hand is described by the general theory of relativity. Because the other fundamental interactions are quantized, it therefore seems natural that in a grand unified theory, a theory of all the fundamental forces, gravity is quantized as well into perhaps Quantum gravity.

A theory of quantum gravity is needed to describe things that are very small but also very heavy, like black holes or the early universe. However, the development of a quantum theory of gravity seems difficult on grounds that, in general relativity all physical qualities have definite values, whereas in quantum mechanics they do not as shown in Heisenberg's uncertainty principle.

The problems in General Relativity arise from trying to deal with a universe that is zero in size (infinite densities). But quantum mechanics suggests that there may be no such thing in nature as a point in space-time, implying that space-time is always smeared out, occupying some minimum region. The minimum smeared-out volume of space-time is a profound property in any quantized theory of gravity and such an outcome lies in a widespread expectation that singularities will be resolved in a quantum theory of gravity.

However, Prof Brian Dolan at the Department of Theoretical Physics, NUI Maynooth, is quick to point out that there is not yet any set agreement on what a theory of quantum gravity should look like, or even on the exact problem it is trying to solve."There is no accepted theory of quantum gravity," he says. "There are currently a number of contenders, and by far the most popular is superstring theory. Many physicists find superstring theory compelling due to its internal elegance, but despite decades of intense research it has not produced a single experimentally testable result." He suspects that trying to unite general relativity and quantum mechanics may be the wrong way to go, and that any future breakthrough may come from a completely unexpected direction; perhaps from some young mind with a fresh perspective.

This chapter employs new idea towards the development of a quantum theory of gravity in a bid to solve the following unsolved problems in physics;

(i) Is it true that at every spatial dimension, there exists new physics and that it is the work of Physicists to find out? What is the method or procedure through which new physics can be found? Does this require extra dimensions?

(ii) Does nature have more than four space-time dimensions? If so, what is their size? Are dimensions a fundamental property of the universe or an emergent result of other physical laws? Can we experimentally observe evidence of higher spatial dimensions?

(iii) Can the singularities that plague the General theory of Relativity be resolved in any quantum theory of Gravity?

The Standard Model is inconsistent with that of general relativity, to the point that one or both theories break down under certain conditions (for example within known spacetime singularities like the Big Bang and the centers of black holes beyond the event horizon).

The appearance of singularities in any physical theory is an indication that something is wrong and that there is a need for new physics. Singularities can be avoided in GR and any field theory through the introduction of an efficient regularization procedure as this book directs.

Regularization is a method of modifying observables which have singularities in order to make them finite by the introduction of a suitable parameter called regulator. The regulator, also known as a "cutoff", models our lack of knowledge about physics at unobserved scales (e.g. scales of small size or large energy levels). **It compensates for the possibility that "new physics" (beyond the SM) may be discovered at those scales which the present theory is unable to model,** while enabling the current theory to give accurate predictions as an "effective theory" within its intended scale of use.

The need for regularization terms in any quantum field theory of quantum gravity is a major motivation for Physics beyond the standard model. Infinities of the non-gravitational forces in QFT can be controlled via renormalization only but additional regularization and hence new physics is required uniquely for gravity. The regularizers model, and work around, the breakdown of QFT at small scales and thus show clearly the need for some other theory to come into play beyond QFT at these scales. A. Zee (Quantum Field Theory in a Nutshell, 2003) considers this to be a benefit of the regularization framework, theories can work well in their intended domains but also contain information about their own limitations and point clearly to where new physics is needed.

Therefore the main objective of this section is to discover new physics at those scales (or extra dimensions) which the General relativity theory and Quantum mechanics is unable to model. The section also sets out to prove that due to quantum gravitational effects, there is a minimum distance beyond which the force of gravity no longer continues to increase (operate) as the distance between the masses become shorter.

General Theory

During the years, strong evidence has appeared that the acceleration of any physical object cannot be arbitrarily large, but it should be superiorly limited. For example in string theory, it was derived that string acceleration must be less than some critical value, determined by the string tension and its mass. From the classical point of view (as Wheeler suggested), if we consider an extended object in **rotating motion**, we have the acceleration $a = v^2/R$ and it follows that a, must be at least limited by c^2/R. However to differ from the classical Newtonian mechanics and

Einstein's General relativity theory we introduce a regulator "Cutoff" $\alpha_g{}^n$, where α_g is the gravitational coupling constant, R is the distance between two masses and n is a positive number (**extra dimension** number), then the acceleration must be limited by $a = \frac{c^2}{2R}\alpha_g{}^n$ (i), (Assuming a diameter of 2R).

Thus to avoid the infinity but while retaining the point nature of the particle would be to postulate a small additional dimension **n** over which the particle could 'spread out' rather than over 3D space.

For example, in the Unruh temperature we can only and only deduce both the Hawking temperature and maximal temperature (Sakharov Temperature) under the assumption of the existence of a maximal acceleration given in formula (i) above as,

The Unruh temperature is given as,

$$T = \frac{\hbar a}{2\pi c k}$$

Since the acceleration is known from (i) above, then the temperature will reduce to,

$$T = \frac{\hbar c}{4\pi R k} \alpha_g^{\,n}$$

For a Schwarzschild Black hole of radius $R = \frac{2GM}{c^2}$, the temperature reduces to

$$T = \frac{\hbar c^3}{8\pi GMk} \alpha_g^{\,n}$$

Since the gravitational coupling constant has a formula $\alpha_g = \frac{GM^2}{\hbar c}$, taking values of n=0,1,2,..............,N. Then the Hawking temperature will become a result of n=0 extra spatial dimensions as,

$$T = \frac{\hbar c^3}{8\pi GMk}.$$

Also the maximum temperature (Sakharov temperature) is deduced at n=1/2 as ,

$$T = \frac{1}{8\pi k}\left(\frac{c^5\hbar}{G}\right)^{1/2}$$

Therefore the temperature of a black hole increases as a black hole loses mass in Hawking Black hole evaporations. The analysis given above is a clear indication that the temperature doesn't increase exponentially as it has been known from Hawking's original proposals, there is a maximum temperature, a limit on temperature that screens (resolves) the classical singularity. It is therefore true that the radiation spectrum contains all Standard Model particles, which are emitted on our brane, as well as gravitons, which are also emitted into the extra dimensions. It is expected that most of the initial energy is emitted during this phase in Standard Model particles. Therefore we recommend the applications of a factor α_g^n in situations involving the examination and experimentation of quantum gravitational phenomenon. We shall see in the coming chapters that such a factor when used in loop quantum cosmology it reproduces both the results of loop quantum gravity and string theory.

The idea of including extra dimensions, to achieve the goal of unifying physics, is not a new one. Already the year before Einstein in 1915 introduced his theory of general relativity; Gunnar Nordstrom suggested a unification of gravity and electromagnetism with the introduction of a fifth dimension. These forces were the two only forms of interaction known at that time. But this idea was forgotten for some time with the eruption of the First World War. But in April 1919 Theodor Kaluza introduced independently, in a letter to Einstein, a fifth dimension in an attempt to unify Einstein's theory of gravity and Maxwell's theory of light. Oskar Klein (1926) contributed, in this quest, with his assumption that the extra dimension was compactified. The Kaluza-Klein theory was a fact. This theory includes an extra space dimension that is rolled up into a tiny circle, i.e. compactified. And in this five dimensional theory, there is only one underlying force, gravity. But in the four-dimensional spacetime observed at great distances, it appears to be three kinds of forces, among these a gravitational and an electromagnetic force. This topic was initially a popular topic for research, but lost much of its interest with the introduction of quantum mechanics.

In recent years the topic of extra dimensions has experienced a renewed interest. This renewed interest is also due to the exciting possibility of observing new and spectacular physical phenomena at far lower energy scales than otherwise. Even at energies available in the not so distant future, these phenomena could appear. Among these is the creation of higher dimensional semi-classical microscopic black holes. The possibility of observing these objects, is viewed as an opportunity to perhaps discover new intriguing physics.

Therefore from (i) using Einstein's equivalence principle we get the minimum distance beyond which the force of gravity no longer continues to increase as;

$$R = \frac{R_s}{\alpha_g{}^n} \qquad \text{(ii)}.$$

Where $R_s = \frac{2GM}{c^2}$ is the Schwarzschild radius. We therefore conclude that;

(i) At n=0 extra spatial dimension, we have a physical theory of General relativity at a length scale of $R = R_s = \frac{2GM}{c^2}$ - the Schwarzschild radius.

(ii) At n=1/2 extra dimension, we have the quantum theory of gravity (New physics) at the Planck length scale $l_p = \sqrt{\frac{\hbar G}{c^3}}$

(iii) At n=1 extra dimension, we have the theory of Quantum mechanics at the Compton wavelength scale of $\lambda = \frac{\hbar}{mc}$.

(iv) Lastly at n=2 we have new physics at a length scale $R = \frac{\hbar^2}{GM m^2}$ and the journey continues.

According to the Standard Model of particle physics, the world is governed by four fundamental forces: gravity, electromagnetism, and the weak and strong nuclear forces. Although things act a bit "spooky" down on the quantum level, science has managed to generally describe all of these forces at both the macro and quantum scales – except gravity.

Gravity is the weakest of the fundamental forces, and it's been suggested that this is because some gravitons (the hypothetical particles) that carry the gravitational force tend to escape into extra dimensions. We're simply too big to travel through or even notice these other dimensions.

So, to study whether these extra dimensions are lurking in extremely tiny spaces, the researchers from Osaka, Kyushu and Nagoya Universities set out to test gravity on the sub nanometer scale. To do so, they used the world's highest intensity neutron beam, which is housed at the Japan Proton Accelerator Research Complex (J-PARC).

The team found that the results matched predictions based on the known laws of physics, which indicates that Newton's law still applies as expected down to a scale of less than 0.1 nanometers. No unexplained force ie, another dimension is acting on these particles at this scale.

That doesn't mean those extra dimensions aren't there, just that they may be hiding at even smaller scales still. The researchers are currently working to further improve the sensitivity of the equipment, which might help them probe those tiny spaces.

In a completely different context, an international team of researchers led by Professor Immanuel Bloch (LMU/MPQ) and Professor Oded Zilberberg (ETH Zürich) has now demonstrated a way to observe physical phenomena proposed to exist in higher-dimensional systems in analogous real-world experiments. Using ultracold atoms trapped in a periodically modulated two-dimensional superlattice potential, the scientists could observe a dynamical version of a novel type of quantum Hall effect that is predicted to occur in four-dimensional systems. (Nature, 4 January 2018)

"Physically, we don't have a 4D spatial system, but we can access 4D quantum Hall physics using this lower-dimensional system because the higher-dimensional system is coded in the complexity of the structure," a researcher with the US-based team, Mikael Rechtsman from Penn State University, told Ryan F. Mandelbaum at Gizmodo. "Maybe we can come up with new physics in the higher dimension and then design devices that take advantage the higher-dimensional physics in lower dimensions."

The above statements can be summed up in the following simplest model;

Let the Gravitational force between two identical particles be related to the magnetic force between them and similarly let the electric force between two particles be related to the magnetic force as;

Gravitational force $(\frac{Gm^2}{R^2})$ = **magnetic force (Bec)** × α_g^n (iii)

and

Electric force $(\frac{e^2}{4\pi\varepsilon R^2})$ = **magnetic force (Bec)** × α_e^n (iv)

Where α_e is the electromagnetic coupling constant- Fine structure constant

The magnetic flux, represented by the symbol **Φ**, threading some contour or loop is defined as the magnetic field **B** multiplied by the loop area, $A=\pi R^2$, i.e. **Φ** = **B** · **A**. Obviously, both **B** and **A** can be arbitrary and so is **Φ**. The inverse of the flux quantum, $1/\Phi_0$, is called the **Josephson constant**, and is denoted K_J.

However, if one deals with the superconducting loop or a hole in a bulk superconductor, it turns out that the magnetic flux threading such a hole/loop is quantized. Therefore the magnetic flux quantum from (iii) and (iv) will be given by,

$$\Phi_G = \pi G m^2 / e c \alpha_g^n$$

$$\Phi_E = e / 4\varepsilon c \alpha_e^n$$

Such that at n=0 extra dimension,

$$\Phi_G = \pi G m^2 / e c$$

$$\Phi_E = e / 4\varepsilon c$$

The above given values represent the classical flux at 3D spatial dimensions.

At n=1/2 extra dimension,

$$\Phi_G = \frac{\pi m}{e} \left(\frac{Gh}{c}\right)^{1/2}$$

$$\Phi_E = \left(\frac{\pi \hbar}{4\varepsilon c}\right)^{1/2}$$

These represent the quantum theory of Gravity.

At n=1 extra dimension,

$$\Phi_G = \pi\hbar/e$$

$$\Phi_E = \pi\hbar/e$$

These represent the magnetic flux quantum at the quantum scale. Also at n=1 the magnetic flux value is the same in both equations, meaning that the gravitational force becomes analogous to the electromagnetic force at n=1.

In other words, just as a 3D object casts a 2D shadow, scientists have managed to observe a 3D shadow potentially cast by a 4D object – even if we can't actually see the 4D object itself. That could unlock some new findings in the very fundamentals of science.

14. How can the Laws of Physics be derived from one Underlying Principle?

In this section a new approach towards Quantum Gravity is presented, we try to study the theory from new assumptions which are far more different from models that have been used by scientists for centuries. Most physicists have clung to old models or complex mathematical scientific methods to explain phenomenon. They are trying to explain physics using the mathematics that was earlier used by Einstein, Richard P. Feyman etc. The mathematical ideas that were presented by these physicists were complicated and such has been difficult to understand and of course misleading (Read Lost in Math, a book by Sabine Hossenfielder). For example, in a statement by Dr Lee Smolin, "the mathematisation of physics has resulted in the reduction of the cosmos to a mathematical entity, which has not only confused physicists but accounts for their worst and most distracting assertions".

There is a wide spread speculation that the mathematical formulation of physics has not only confused physicists but has also lead to failures in the development of a quantum theory of gravity.

Physics as a subject should be simple and elegant, trying to explain everything from one source. In other words trying to explain all of physics from one equation call it "**the principle of least action**". Imagine deducing the equations of gravity, quantum mechanics, electromagnetism, heat etc from one equation, wouldn't it be unique than holding about ten books about a different subject of physics each starting from its own source?

The principle of least action in simple terms means; to understand how to get from point A to point B using the least amount of physical work for example taking an elevator rather than using the stairs, in otherwords deducing the most fundamental physical equations from one principle as we are yet to find out.

Assuming that the ratio of the gravitational force to the electric force is equal to the gravitational coupling constant, we then have,

$$[8\pi G/c^4][Ee][GM^2/c]=n^2\hbar \qquad 1$$

Where G– gravitational constant, n- quantum number, c- constant speed of light, e- charge on an electron, M – mass and is the reduced Planck constant. **Ee** is the electromagnetic force, and **8πG/ c⁴** is the gravitational force at the swcharzichild's radius. From the above given principle, we deduce the temperature of a black hole, the time taken by a black hole to evaporate, entropy of a black hole, the wiedemann franz law and the stefan's radiation law.

1. The temperature of a black hole

On arranging equation one to get the random translational kinetic energy, we obtain

$$[GM/c^2]Ee = n^2 c^3 \hbar /8\pi\, GM = kT$$

Where k is the boltzmann's constant and T is the temperature. Hence at n=1,

$$T = c^3 \hbar /8\pi\, GMk \qquad\qquad 2$$

This is the known temperature of a black hole that was originally derived by Hawking

2. Time taken by a black hole to evaporate

On dividing through eqn1 by the momentum Mc we obtain, the time t given by

$$t = Mc/Ee = 8\pi G^2 M^3 / n^2 \hbar c^4$$

Such that when n=0.03953

$$t = 5120\pi G^2 m_o^3 / \hbar c^4 \qquad 3$$

This is the known time of a black hole that was originally derived by Hawking

3. Entropy of a black hole

Squaring both sides of equation 1 and arranging we generate the intensity as

$$W/tA = E^2e^2/2nh = n^3c^{10}\hbar/256\pi^3 G^4M^4 \qquad 4$$

Where A is the area on which the radiations fall, W is energy, and t is time. But entropy is energy divided by temperature Eq so then

$$W/T = S = (n^3 c^{10} \hbar/256\pi^3 G^4M^4)(tA/T)$$

Since t is known from Eq3 and T from Eq2, then at $n = \pi$

$$S = Akc^3/4G\hbar$$

This is the known Bekenstein-Hawking area entropy law

4. Thermal properties of solids

From the intensity equation4,

$$E^2e^2/2nh = c^{10}\hbar/256\pi^3 G^4M^4$$

Arranging the above equation to introduce in the transilational kinetic energy obtained above [$kT= c^3\hbar/8\pi GM$], we have

$$\pi M^2G^2E^2/3c^4 =(\pi^2/3e^2)(c^3\hbar/8\pi GM)^2=(\pi^2/3e^2)T^2k^2$$

Dividing both sides by T we obtain on the left hand side of the equation the ratio of thermal conductivity K to electric conductivity δ as

$$K/\delta =(\pi^2/3)(k/e)^2 T$$

This is the known wiedemann fanz law

5. Stefan's Radiation Law

Still from the intensity Eqn4 we can arrange the expression on the left hand side of the equation, to read as

$$W/tA = E^2e^2/2nh = (1.875n^3/\pi^2)(\pi^2/60h^3c^2)(c^3\hbar/8\pi GM)_4$$

This is the same as Eqn4 only that it is arranged to predict something. But at n=1 and $(c^3 \hbar /8\pi GM) = kT$, the rate at which energy is radiated is given by

$$W/t = A(1.875/\pi^2)(\pi^2/60h^3c^2)(Tk)^4 = 0.19\sigma AT^4$$

Where $\sigma = \pi^2/60h^3c^2$ is the Stefan boltzmann's constant

In conclusion, I encourage further research into this field. In other words this could be a stepping stone towards the development of a theory of everything via a simpler path of least action.

15. Is Gravity and the Laws of Physics Emergent?

Starting from first principles and general assumptions we present a heuristic argument that shows that Newton's law of gravitation and Coloumb's law of electricity naturally arise in a theory in which space emerges through a zero- point fluctuation of the quantum vacuum. Gravity is identified with a casimir force caused by quantum vacuum fluctuations due to the presence of material bodies in it or the distortion of the vacuum through its interaction with mass. A relativistic generalization of the presented arguments directly leads to the Einstein equations. When space is emergent even Newton's law of inertia needs to be explained. The equivalence principle suggests that it is actually the law of inertia whose origin is casimir.

The real origin of gravity is one of the most important, complex and substantially yet unsolved questions in Physics. The replacement of the Newtonian model of gravity with the Einstein's one given by General Relativity (GR) has only shifted the question without solving it. Within GR, gravity has two possible interpretations: a field one and a geometric one. According to the latter, that has become the prevalent one, gravity is due to the curvature of the space – time "tissue", represented as a "rubber sheet", due to the presence of a mass. Nevertheless, this is a purely mathematical description telling nothing about the physical mechanism starting the motion. In fact, even supposing the existence, in the neighbouring of a source mass, of a curved four – dimensional manifold it doesn't explain why a second particle at rest should move towards the source mass.

As such, it invites attempts at derivation from a more fundamental set of underlying assumptions, and six such attempts are outlined in the standard reference book Gravitation, by Misner, Thorne, and Wheeler (MTW). ' Of the six approaches presented in MTW, perhaps the most far-reaching in its implications for an underlying model is one due to Sakharov; namely, *that gravitation is not a fundamental interaction at all, but rather an induced effect brought about by changes in the quantum fluctuation energy of the vacuum when matter is present.* ' In this view the attractive gravitational force is more akin to the induced van der Waals and Casimir forces, than to the fundamental Coulomb force. Although speculative when first introduced by Sakharov in 1967, this hypothesis has led to a rich and ongoing literature on quantum-fluctuation-induced gravity that continues to be of interest. In this approach the presence of matter in the vacuum is taken to constitute a kind of set of boundaries as in a generalized Casimir effect, and the question of how quantum fluctuations of the vacuum under these circumstances can lead to an action and metric that reproduce Einstein gravity has been addressed from several viewpoints.

Therefore in this chapter we want to show that gravitation might be not a fundamental interaction but a byproduct of the electromagnetic interaction, precisely an electromagnetic phenomena induced by the presence of matter in the quantum vacuum (the quantum field that is present even in empty space). Which means that, matter is not just there but is in the quantum vacuum, and therefore interacts with it, causing some kind of quantum fluctuation energy, that fluctuation is gravitation. In simple terms, a body immersed in quantum fields will interact with them causing gravity to manifest.

(a)Emergence of the laws of Newton

Haisch, Rueda, and others have made the claim that the origin of inertial reaction forces can be explained as the interaction of electrically charged elementary particles with the vacuum electromagnetic zero-point field expected on the basis of quantum field theory.

Gravity is treated as a residuum force in the manner of casimir or vander waals forces. Expressed in the most rudimentary way this can be viewed as follows. The zero point field causes a given charged particle to oscillate. Such oscillations give rise to secondary electromagnetic fields. An adjacent charged particle will thus experience both the zero point field driving forces causing it to oscillate, and in addition forces due to the secondary fields produced by the zero point field driven oscillations of the first particle. Similarly, the zero point field driven oscillations of the second particle will cause their own secondary fields acting back upon the first particle. The net effect is an attractive force between the particles.

Force and Inertia

For the interaction between two particles, each mass experiences a background zero point field and a zero point driven dipole field of the other mass.

Two masses A and D (taken here to be equal for ease of discussion) with D located a distance R from A, along the positive z axis of a coordinate system centered at A. The zero point field will cause a charged particle A to oscillate. The oscillations will then give rise to a secondary electromagnetic field , which will cause particle D to oscillate. In the same way, the zero point field driven oscillations of particle D will cause their own secondary fields acting back upon particle A. the net effect will be an attractive force between particles A and D that will cause one to move towards the other with a small acceleration a_0 in the weak field limit.

Analogous to the Compton Effect, the wavelength of the electromagnetic waves emitted or scattered as a result of particle A interacting with the quantum vacuum will be given as,

$$\lambda_1 = \frac{2\pi R^2}{\lambda_c}\left(\frac{B_0}{B}\right)\alpha$$

Where,

B_0- is the strong magnetic field (greater than or equal to the critical value, which can create electron-positron pairs from the quantum vacuum). The Schwinger mechanism has two cornerstones, the first one is the existence of quantum vacuum and the second one the existence of an external electric field (which attempts to separate electrons and positrons). There are no particles in the vacuum (in that sense the vacuum is empty); but the vacuum is plenty of short-living virtual particle-antiparticle pairs which in permanence appear and disappear (allowed by time- energy uncertainty relation). A "virtual" pair can be converted into a real electron-positron pair only in the presence of a strong external field, which can spatially separate electrons and positrons, by pushing them in opposite directions, as it does an electric field. Therefore the zero point field or quantum vacuum exists but with an external magnetic field stronger than the critical value such that when a particle A is immersed in this zero point field or quantum vacuum, it will interact with the quantum vacuum causing quantum vacuum fluctuations which will trigger the external magnetic field causing oscillation of particle A giving rise to secondary electromagnetic fields.

B- is the value of the magnetic field that exists between particle A and D. This value depends only on the masses of the particles. In other words it is the magnetic field that depends on the matter constituents of the particles in question irrespective of the distance.

$$B = \frac{m^2 ec}{4\pi\varepsilon_o \hbar^2}$$

λ_c- is the reduced Compton wavelength $\frac{\hbar}{mc}$ and α is a dimensionless coupling constant.

It must be noted that, in the weak field limit, the resistance which defines the inertia of a particle is, ultimately, electromagnetic resistance caused by the zero point field on the particle, and it is this resistance which produces gravitational waves with a wavelength of due to a state of motion of a particle,

$$\lambda_2 = \frac{2\pi c^2}{a_o}\alpha$$

From the above given assumption, it is proposed that a body's inertia is due, to the distribution of matter in the universe, and, more precisely, to the electromagnetic interaction that arises from quantum fluctuations of the zero point field in accelerated frames. Basically, a particle's inertia is a function of the particle's interaction with zero point field. Inertia is resistance to acceleration and this reistance causes a form of the gravitational wave simply because resistance becomes a force. This implies that, the resistance which defines the inertia of a particle is, ultimately, electromagnetic resistance caused by the zero point field on the particle.

It must therefore be true that under a condition where, $\lambda_1 = \lambda_2$, we recover Newton's law of inertia (F=ma) as,

$$F = ma_o = \frac{\hbar c}{R^2}\left(\frac{B}{B_o}\right) \qquad (1)$$

Therefore matter continuosly interacts with the zero point field (as Casimir effect), and this interaction yields a force (the resistance to motion) whenever acceleration takes place. Inertia is due to the distortion of the zero point fluctuations in an accelerated reference frame. Technically, inertia is due to the high frequencies of the distortion of the zero point spectrum.

Newton's law of gravity

For the interaction between two masses, each mass experiences a background zero point field and a zero point field driven dipole field of the other mass. The procedure followed here is precisely that developed by Boyer for the derivation of the retarded van der waals forces at all distances between a pair of polarizable particles. Therefore we need only outline the procedure as it applies to our case.

Two masses A and D (taken here to be equal for ease of discussion) with D located a distance R from A, along the positive z axis of a coordinate system centered at A. The modified casimir force between the pair of particles A and D is given by Eqn1,

$$F = ma_o = \frac{\hbar c}{R^2}\left(\frac{B}{B_o}\right)$$

Where B_o is the external (or background) magnetic field stronger than the critical value and B is the dipole magnetic field at the position of particle A due to the motion of particle D and so forth. But since, $B_o = \frac{ec^2}{4\pi\varepsilon_o G\hbar}$, and $B = \frac{m^2 ec}{4\pi\varepsilon_o \hbar^2}$, on substitution into Eqn1 we obtain a familiar law,

$$F = \frac{Gm^2}{R^2}$$

We have recovered Newton's law of gravitation, practically from first principles!

These equations do not just come out by accident. It had to work, partly for dimensional reasons. In a sense we have reversed these arguments. But the logic is clearly different, and sheds new light on the origin of gravity: it is a casimir force! That is the main statement, which is new and has not been made before. If true, this should have profound consequences.

It is hereby proposed that, gravity is not a separately existing fundamental force, but rather a residuum force derived from zero-point fluctuations of other fields in the manner of the Casimir and van der Waals forces. Particularizing this hypothesis to the zero point fluctuation of the vacuum electromagnetic field, we identify the gravitational force as the casimir force associated with the long-range radiation fields (as opposed to the usual shorter-range induction fields) generated by the particle motion response to the zero point fluctuation of the electromagnetic field.

It is therefore seen that a well-defined, precise quantitative argument can be made that gravity is a form of long-range casimir force associated with particle response to the zero-point fluctuations of the electromagnetic field. As such, the gravitational interaction takes its place alongside the short-range van der Waals forces and the Casimir force as related phenomena which emerge from the underlying dynamics of the interaction of particles with the zero-point auctuations of the vacuum electromagnetic field.

(b)Emergence of electromagnetism

Electromagnetism or the coloumb force emerges in a similar fashion as the gravitational force. The origin of the electric force here assumes a critical magnetic field (Schwinger effect or limit) taken here to represent the external magnetic field,

$$B_o = \frac{m^2 c^2}{\hbar e}$$

But since, $= \frac{m^2 ec}{4\pi\varepsilon_0 \hbar^2}$. On substitution into Eqn1 we obtain a familiar law,

$$F = \frac{e^2}{4\pi\varepsilon_o R^2}$$

We have recovered Coulomb's law of electricity, practically from first principles!

These equations do not just come out by accident. It had to work, partly for dimensional reasons. In a sense we have reversed these arguments. But the logic is clearly different, and sheds new light on the origin of electricity and gravity: it is a casimir force!

Therefore the gravitational field is the set of all electromagnetic fields generated by all particles as they interact with the zero point field. Gravity and electricity results from a distortion of the quantum vacuum through its interaction with a mass.

A more Simpler Derivation: Emergence of Gravity

The first attempt into the derivation of the gravitational force and Newton's laws of inertia was given in part by Erik Verlinde (2011) in which he stated that gravity is an entropic force. Simple as it was, his ideas were on a large scale rejected by mainstream physicists. The rejection of verlinde ideas where not backed up by another approach as it has been with other theories save for Sabine Hossenfelder approaches and critics. I think Erik was not surprised by these attacks because this is what physicists do especially when ones analysis or derivation doesn't involve the use of rigorous mathematical models-the one which were used by Einsten and others.

Anyway, what could be another approach towards the derivation of the gravitational, electricity and the law of inertia different from Verlindes idea of the emergent of gravity as an entropic force?

In this chapter we derive Newton's laws of inertia, gravitation and also the electromagnetic force law from first principles without assuming dark matter and the MOND theories. To differ from Verlindes approach we shall use the notions of Quantum vacuum and the Schwinger effect or limit in QED.

In order to understand the physical significance of the derivation to be given herein, we must remember the Schwinger mechanism (Schwinger, 1951) in Quantum Electrodynamics: a strong electric field, greater than a critical value, can create electron-positron pairs from the quantum vacuum.

The Schwinger mechanism has two cornerstones, the first one is the existence of quantum vacuum and the second one the existence of an external electric field (which attempts to separate electrons and positrons). There are no particles in the vacuum (in that sense the vacuum is empty); but the vacuum is plenty of short-living virtual particle-antiparticle pairs which in permanence appear and disappear (allowed by time- energy uncertainty relation). In simple words, the quantum vacuum is a kingdom of the virtual particle-antiparticle pairs; a kingdom with apparently perfect symmetry between virtual matter and virtual antimatter.

A "virtual" pair can be converted into a real electron-positron pair only in the presence of a strong external field, which can spatially separate electrons and positrons, by pushing them in opposite directions, as it does an electric field. Thus, "virtual" pairs are spatially separated and converted into real pairs by the expenditure of the external field energy. For this to become possible, the potential energy has to vary by an amount in the range of about one Compton wavelength, which leads to the conclusion that a significant pair creation occurs only in a very strong external field E, greater than the critical value.

Therefore the external force which attempts to separate particles and antiparticles converting a virtual pair into a real one may be simplified as,

$$F\Phi = 4\pi E_o \hbar \qquad (a)$$

Where F is the external force, Φ is the magnetic flux impeding a sphere of radius R and area $A = 4\pi R^2$ and E_o is the electric field on an electron of mass m

Derivation of force and inertia

The magnetic field is related to the electric field by

$$E_o = B_o c = \frac{Mmec^2}{4\pi\varepsilon_o \hbar^2} \qquad (b)$$

Remember the above given value is constant for a particle with mass m. There is also an assumption for M=m. Note also that $B_o \neq B$

A new assumption and probably the most surprising one is that, the magnetic flux can be related to energy W and acceleration a by

$$\Phi = \frac{We}{a\hbar\varepsilon_0} \quad \text{(c)}$$

Thus, "virtual" pairs are spatially separated and converted into real pairs by the expenditure of the external field energy. For this to become possible, the potential energy has to vary by an amount in the range of about one Compton wavelength , which leads to the conclusion that a significant pair creation occurs only in a very strong external field E, greater than the critical value E_o.

But, $W = Mc^2$, it is therefore evident why equation (c) for the magnetic flux was chosen to be of the given form. It is picked precisely in such a way that one recovers the second law of Newton

F=ma

As easily verified by combining (a) together with (b) and (c)

Therefore, a similarity or resemblance between acceleration, thermodynamics and electromagnetism comes alive in the statement below;

As there is a formula for the temperature T that is required to cause an acceleration equal to a, $T = \frac{\hbar a}{2\pi k c}$ so there must also be a temperature required to cause an acceleration for an electron in the quantum vacuum at a constant electric and magnetic flux, $T = \frac{\hbar a \varepsilon_o \Phi}{k e}$

Derivation of Newton's law of Gravity

Suppose our universe is a sphere of area $A = 4\pi R^2$ with a sea of virtual particles in a quantum vacuum. It is theorized as before that to separate the virtual particle and antiparticles in a vacuum into real particles one will require a strong external electric or magnetic field B. For the gravitational field, this external magnetic field was calculated to be,

$$B = \frac{e c^2}{4\pi G \hbar \varepsilon_o}$$

Where we introduced a new constant G. Eventually this constant is going to be identified with Newton's constant, of course. But since we have not assumed anything yet about the existence of a gravitational force, one can simply regard this equation as the definition of G.

Then, the magnetic flux will be given as,

$$\Phi = BA = \frac{ec^2 R^2}{G\hbar\varepsilon_0} \quad \text{(d)}$$

Substituting Equation (b) and (d) into (a) one obtains the familiar law

$$F = G\frac{Mm}{R^2}$$

We have recovered Newton's law of gravitation, practically form first principles. Following the above derivation carefully, it implies that gravity is a quantum force resulting from the quantum fluctuations of the vacuum due to an existence of an external strong electric or magnetic field separating particles from antiparticles or matter from anti matter.

Derivation of the law of electromagnetism

Following the same steps as in the previous derivation for the gravitational force but assuming a different external magnetic field (Schwinger limit),

$$B = \frac{Mmc^2}{\hbar e}$$

Then, the magnetic flux will be given as,

$$\Phi = BA = \frac{4\pi Mmc^2 R^2}{\hbar e} \quad (e)$$

Substituting Equation (e) and (b) into (a) one obtains the familiar law

$$F = \frac{e^2}{4\pi\varepsilon_0 R^2}$$

We have recovered Coulomb's law of electromagnetism practically form first principles.Therefore in principle, every external force which attempts to separate particles and antiparticles, may convert a virtual pair into a real one. If it is always an attractive force, as commonly believed today, gravity can't separate particles and antiparticles. Hence, the conjectured gravitational repulsion between matter and antimatter is a necessary condition for separation of particles and antiparticles by a gravitational field and consequently for the creation of particle-antiparticle pairs from the quantum vacuum. But while an electric field can separate only charged particles, gravitation as a universal interaction might create particle-antiparticle pairs of both charged and neutral particles. Thus, the hypothesis of antigravity opens possibility for a gravitational version of the Schwinger mechanism.

In conclusion, gravity is not an entropic force but rather a quantum force stemming from the quantum fluctuation of particles in the vacuum. That is the main statement, which is new and has not been made before. If true, this should have profound consequences.

16. The Galaxy Rotation Problem

The Big problem*: Is dark matter responsible for differences in observed and theoretical speed of stars revolving around the centre of galaxies, or is it something else?*

Why would we want to modify Einstein's outstanding intellectual achievement?

a) Newtonian and Einstein gravity cannot describe the motion of the outermost stars and gas in galaxies correctly.

b) If dark matter is not detected and does not exist, then Einstein's and Newton's gravity theories must be modified.

Since the 1970s and early 1980s, a growing amount of observational data has been accumulating that shows that Newtonian and Einstein gravity cannot describe the motion of the outermost stars and gas in galaxies correctly, if only their visible mass is accounted for in the gravitational field equations.

To save Einstein's and Newton's theories, many physicists and astronomers have postulated that there must exist a large amount of "dark matter" in galaxies and also clusters of galaxies that could strengthen the pull of gravity and lead to an agreement of the theories with the data. This invisible and undetected matter removes any need to modify Newton's and Einstein's gravitational theories. Invoking dark matter is a less radical, less scary alternative for most physicists than inventing a new theory of gravity.

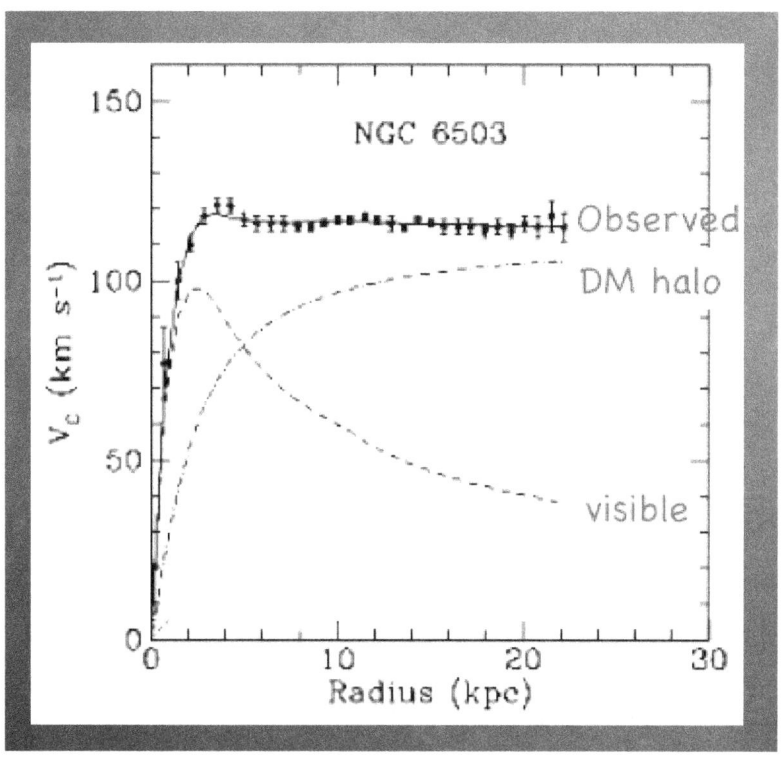

Fig. Galaxy data that show that Newtonian and Einstein gravity do not fit the observed speed of stars in orbits inside a galaxy such as NGC 6503

If dark matter is not detected and does not exist, then Einstein's and Newton's gravity theories must be modified. Can this be done successfully? Yes! My modified gravity

(MOG) can explain the astrophysical, astronomical and cosmological data without dark matter.

(a)The Modified Force Law

Let the force of gravity be simplified as,

$$F = p\sqrt{\frac{2\pi\alpha g}{\lambda}} \qquad (1)$$

Where p=mc, is the relativistic momentum of a particle of mass m. α is the coupling constant which determines the strength of a force in any interaction at the range determined by the Compton wavelength λ. $g = \frac{GM}{R^2}$ is the acceleration due to gravity for a point particle at a distance R from a star of mass M.

The above given law was used in chapter 10 and chapter 11 and it was expressed as, $F = \frac{me}{R}\sqrt{\frac{GM\omega}{4\pi\hbar\varepsilon_0}}$

Where ω=2πf is the angular frequency of the graviton-photon oscillations and e is the charge on an electron. In a limit of $\omega = \frac{GM}{R^2}\left(\frac{4\pi\hbar\varepsilon_0}{e^2}\right) = \frac{g_N}{c\alpha_e}$, where g_N is the usual Newtonian acceleration *due to gravity and α_e is the fine structure constant, the above new force law reduces to the Newtonian law of universal gravitation.*

It must be noted that the above force law Eqn1 reduces to the Newtonian law of gravity $\frac{GMm}{R^2}$, only when the following two conditions are met:-

Condition1. The wavelength is equal to the circumference of the circle swept out by the orbiting mass m around M, that is $\lambda=2\pi R$.

Condition 2. When the coupling constant is half the Newtonian deflection angle θ for a light ray under the influence of a high gravitational field at the sun's surface ($\alpha = \theta/2 = R_s/2R$, where Rs is the Schwarzichilds radius)

When the two conditions given above are met, then Eqn1 will become the Newton's law of gravity.

(b)The Tully-Fisher Relation

One of the best fit predictions of MOND is a single universal Tully-Fisher relation.

" The relation between asymptotic velocity and the mass of the galaxy is an absolute one" (Milgrom 1983). This is given by, $V^4 = a_o GM$, where $a_o = 1.2 \times 10^{-10} ms^{-2}$. In this chapter an equation similar to the Tully-Fisher relation is deduced from (1) as given below,

For circular orbits about a mass M, we have the centripetal force equal to Eqn1 as,

$$\frac{mV^2}{R} = p\sqrt{\frac{2\pi\alpha g}{\lambda}} = \frac{mc}{R}\sqrt{\frac{2\pi\alpha GM}{\lambda}}$$

This gives an asymptotically rotation velocity independent of R:

$$V^4 = \left(\frac{2\pi\alpha c^2}{\lambda}\right) GM = a_0 GM \qquad (2)$$

It is this behavior that gives rise to *asymptotically flat rotation curves* and the Tully-Fisher relation (Tully & Fisher 1977) without invoking dark matter.

Comparing (2) to the Tully-Fisher relation, we determine the acceleration limit as,

$$a_0 = \frac{2\pi \alpha c^2}{\lambda}$$

From which the coupling constant takes on values of,

$$\alpha = \frac{a_0 \left(\frac{\hbar}{mc}\right)}{2\pi c^2}$$

This result implies that, if the milgrom acceleration was really a constant of $a_o = 1.2 \times 10^{-10} ms^{-2}$ and also the fine structure constant was the coupling constant of 1/137, then it will be true that the wavelength or the range of the interaction will be given by exactly , $\lambda = 3.44 \times 10^{25} m$ which is within the acceptable size of the galaxies. This could help us connect quantum mechanics with gravity at small accelerations.

We anticipate that MOG will modify how stars collapse and the nature of black holes. We know that a supermassive object with mass ~ 3 X 10^6 MSUN is at the center of our

Galaxy (MILKY WAY). We are not able to determine yet whether the object is a GR black hole with a horizon. Perhaps future telescopes and space missions will be able to get close enough to the supermassive object to tell whether it is a black hole in spacetime or some other kind of object. However, as distant observers, we can never see a black hole event horizon form! The formation of the event horizon occurs in the infinite future, so we cannot actually ever see a black hole event horizon form as a star collapses.

(c) Modification of the Newtonian Dynamics (MOND)

From equation (c) Chapter13 we showed that the gravitational acceleration of a particle is related to the kinetic energy W and magnetic flux Φ by the following equation,

$$a_g = \frac{Wq}{\Phi \hbar \varepsilon_0}$$

The above equation implies that, keeping the energy a constant, the acceleration of a particle increases with a fall in the magnetic flux and vice versa is true. Thus the acceleration of a particle is to a great extent affected by the magnetic field.

From the above given relationship we deduce a modification of the Newtonian dynamics in a limit of small accelerations similar to the Milgrom hypothesis (Astrophys. J. 270, 365-1983) but with an electric force as a cause of these small accelerations and we deduce the value of this acceleration in relation to the movement of electrons in the hydrogen atom. Using the same analysis we also deduce the value of the total mass of the universe.

If we let the energy or the kinetic energy, the work done to move the particle around a body of mass M (a galaxy) in the magnetic field created by M, in a process of magnetic induction be given by, $W = mv^2$, where v is the velocity of the particle and m is the mass of the particle. Also if we let the magnetic flux at any point in space at a distance R from M be $\Phi = 4\pi R^2 B$, then the acceleration is simplified in this way,

$$a_g = \frac{mv^2 q}{4\pi R^2 B \hbar \varepsilon_o}$$

This can then be re-written to resemble the Milogram formula in this way,

We know the magnetic field B is related to the velocity and electric field E by, Bv=E, then the acceleration of a particle is,

$$a_g = m\left(\frac{qa}{4\pi\varepsilon_0 E\hbar v}\right)a$$

Where, $a = \frac{v^2}{R}$. This then reduces to,

$$a_g = \left(\frac{q^2 a}{4\pi\varepsilon_0 \hbar v \left(\frac{Eq}{m}\right)}\right)a$$

This then gives,

$$a_g = \left(\frac{a}{\frac{1}{\alpha_e}\left(\frac{Eq}{m}\right)}\right)a$$

From the above given given equation, $\alpha_e = \frac{q^2}{4\pi\varepsilon_0 \hbar v}$ is the fine structure constant. Eq is the electric force on the particle of charge q. Then the acceleration of a particle of mass m in the electric field will be given by, $a_1 = \frac{Eq}{m}$.

Now relating this to the Milogram Hypothesis of $a_g = \left(\frac{a}{a_o}\right)a$, where $a_o \sim 1.2 \times 10^{-10} m/s^2$, we have

$$\left(\frac{a}{\frac{1}{\alpha_e}a_1}\right)a = \left(\frac{a}{a_o}\right)a$$

Therefore in a limit where v=c (the speed of light) and q=e (the charge on an electron), we have

$$\alpha_e = \frac{e^2}{4\pi\varepsilon_o \hbar c} = 1/137$$

This implies that, the acceleration of a particle in an electric field of a body of mass M will be given as

$$a_1 = \alpha_e a_o = 8.759 \times 10^{-13} m/s^2 \quad \text{(d)}$$

The above given value implies that, in the presence of the electric and magnetic field the Milogram acceleration constant must be corrected to fit the data very well.

Then the modified Newtonian law of gravity is written as,

$$F = \frac{GMm}{\left(\frac{\alpha_e a}{a_1}\right)R^2}$$

In a limit $a_1 = a$ we don't recover the Newtonian law except for $\alpha_e = 1$, which proves to be difficult. This therefore requires us to rethink gravity with a new gravitational constant that could read as $G_B = \frac{G}{\alpha_e} = 9.138 \times 10^{-09} Nm^2/kg^2$

(d) Calculation Of The Total Mass Of The Galaxy From Equation (d)

Assuming the ratio of the accelerations to be equal to the fine structure constant as,

$$\frac{g_b}{g_a} = \alpha_e \quad (e)$$

Where,

$$g_a = \frac{8\pi\varepsilon_o \hbar c^5}{e^2 GM}$$

And,

$$g_b = \frac{c^4}{GM}$$

Relating these accelerations to the ones given in equation d we should be able to deduce the mass of the galaxy as,

$$g_a = \frac{8\pi\varepsilon_o \hbar c^5}{e^2 GM} = a_o = 1.2 \times 10^{-10} m/s^2$$

From which the total galaxy mass is given as,

$$e^2 M = 7.122 \times 10^{18} kg C^2$$

For $e = 1.602 \times 10^{-19} C$, the total mass is given as,

$$M = 5.55 \times 10^{56} kg$$

This is almost the total mass of the Universe. *Note: since the total mass of ordinary matter in the universe is known to be* $M = 1.5 \times 10^{53} kg$, *then using the above given equation, the charge required to give this value is* $e = 6.891 \times 10^{-18} C$

But with,

$$g_b = \frac{c^4}{GM} = \alpha_e a_o = 8.759 \times 10^{-13} m/s^2$$

We have,

$$M = 1.387 \times 10^{56} kg$$

Because this is not the same mass obtained with the use of g_a, we then rewrite the equation of g_b to give the same value of mass as,

$$g_b = \frac{nc^4}{GM}$$

Where, n is the number taking on values from, n=2,4,16,.............

In conclusion, we have not only modified the Newtonian dynamics as a requirement to account for darkmatter but we have just deduced the total mass of the universe. This implies that we are near or probably we have just completed the theory of quantum gravity.

17. Particle Creation by Black Holes: Is it Hawking's Approach or My Approach?

In 1975 Hawking calculated quantum mechanically that a black hole will emit particles as if it were a black body with a temperature proportional to its surface gravity. Although this thermal emission is insignificant for black holes formed by stellar collapse, it is of crucial importance for the small primordial black holes formed by density fluctuations in the early universe.

The most significant consequence of a black hole is that, the temperature of a black hole increases as a black hole loses mass. The temperature increases exponentially into a burst of gamma rays leaving a black hole remnant. There is no clear account on this, not until we have fully developed a consistent quantum theory of gravity (where the mass of a black hole approaches the Planck scale of mass and radius).

The evaporation of a black hole starts with a spin down phase in which the Hawking radiation carries away the angular momentum, after which it proceeds with emission of thermally distributed quanta until the black hole reaches the Planck mass.

The radiation spectrum contains all Standard Model particles, which are emitted on our brane, as well as gravitons, which are also emitted into the extra dimensions. It is expected that most of the initial energy is emitted during this phase in Standard Model particles.

One of the major problem with black holes is that, we cannot directly measure any properties of them neither can we produce black holes in any terrestrial experiment. According to Cheung (2002), this is due to the fact that in order to produce black holes in collider experiments one needs a centre of mass energy above the Planck scale, which is obviously inaccessible at the moment. But thanks to the introduction of the numerical coefficient, we can now as this section directs, detect a vast number of black holes in our galaxy by observing and detecting the mass scale or the low energy scale quanta emitted whenever a black hole evaporates due to stellar collapse.

The numerical coefficient α depends on which particle species can be emitted at a significant rate and can be determined by taking the effect of the absorption cross section. This coefficient is of great importance in the standard model and if described in detail it could unlock the secrets hidden deep in the cosmos.

The dominant contribution to α in the standard model comes from fermions, the contribution to α for electrons and positrons is 1.575×10^{-4} (Don Page 1975).

Page calculated the emission rates for massless particles, predicted the lifetime of black holes (from the total power emitted in all modes) and also deduced the numerical coefficient for the dimensionally determined quantities (in terms of the Planck mass etc).The coefficient appears in Eqn25 and Eqn26 of the rate of change of mass and the life time of a black hole by Don Page (1975). There is no known formula relating the numerical coefficient to the mass scale or low energy scale quanta, the mass of an electron and the mass of a proton. Yet this could provide a unique probe of at least four areas of physics: the early Universe; gravitational collapse; high energy physics; and quantum gravity.

Assuming that a black hole emits particles at a mass scale M_* (low energy scale quanta), we propose the Numerical coeffient to be

$$\alpha = \frac{2M_* m_e}{m_p^2} \quad (14)$$

Where, m_p is the mass of the Proton and m_e is the mass of an electron. This concept will help us understand the type of particles emitted by Black holes and how detecting them will help us observe most of the Black holes in space, the coeffient will be efficient in the derivation of the Chandrasekher mass limit and the Bekenstein-Hawking area entropy law. Also this coefficient will be of good use in understanding the lowest energy level of the White dwarf Hydrogen atom. In the table below, we give values of α for different mass scales M_*.

Table 1

α	Mass scale M_* (kg)	Remarks
2.837×10^{15}	4.343×10^{-9}	Planck particle-Planck scale
1	1.531×10^{-24}	Yet to be found≈ 1TeV
0.073	2.23×10^{-25}	Higgs boson
1.575×10^{-4}	2.41×10^{-28}	Pion-Neutral- cosmic rays

Note: M_{pl} (the Planck mass of $\left(\frac{\hbar c}{8\pi G}\right)^{1/2}$ =4.343 × 10^{-9}kg)

From the analysis given above, a black hole of mass M_{BH} will have a temperature and a life time given by

Temperature: $T = \dfrac{M_{pl}^2 c^2}{k M_{BH}} \left(\dfrac{2 M_* m_e}{m_p^2} \right) = \dfrac{M_{pl}^2 c^2}{k M_{BH}} \alpha$

$$T = 8.0305 \times 10^{46} \dfrac{M_*}{M_{BH}}$$

Lifetime: $\tau = \dfrac{G M_{BH}^3 m_p^2}{M_{pl}^2 c^3 M_* m_e} = \dfrac{2 G M_{BH}^3}{M_{pl}^2 c^3 \alpha}$

$$\tau = 8.019 \times 10^{-43} \dfrac{M_{BH}^3}{M_*}$$

Note: The power of a black hole is given by, $P = \dfrac{M_{pl}^2 c^5}{2 G M_{BH}} \alpha$

For purposes of this study, let us limit ourselves to two Primordial Black holes, one with a mass of 4.7×10^{11} kg and another with mass 1.331×10^{17} kg. We calculate the

218

Temperature and life time of these black holes at known and assumed mass scales as given in table 2 and table 3.

Table 2.

Black hole (Kg)	Mass scale M_*	Temp- T(K)	τ (sec)	Remarks
4.7 × 10¹¹	4.343×10^{-9}	7.42×10^{26}	19.17	Early universe
	1.531×10^{-24}	2.62×10^{11}	5.44×10^{16}	Current Age of the Universe

2.23 × 10⁻²⁵	3.81 × 10¹⁰	3.73 × 10¹⁷	Current Age of the Universe
2.41 × 10⁻²⁸	4.12 × 10⁷	3.46 × 10²⁰	?

Note: For $M_* = M_{BH}$ we obtain the temperature of a Black hole $T = 8.0305 \times 10^{46}$ K. This is the maximum temperature of a black hole above which the black hole cease to exist.

Table 3.

Black hole (Kg)	Mass scale M_*	Temp- T(K)	τ (sec)	Remarks

1.33 × 10¹⁷	4.343 × 10⁻⁹	2.62 × 10²¹	4.34 × 10¹⁷	Current Age of the Universe
	1.53 × 10⁻²⁴	9.24 × 10⁵	1.23 × 10³³	
	2.23 × 10⁻²⁵	1.35 × 10⁵	8.46 × 10³³	
	2.41 × 10⁻²⁸	145.52	7.83 × 10³⁶	

We learn from the above tables that, the temperature and lifetime associated with a black hole will not only depend on the mass of a black hole but also on the mass scale of the quanta emitted as scaled from the numerical coefficient which depends on which particle species can be emitted at a significant rate. For example, the theory of black hole radiations that was developed by S.W. Hawking will only become correct and deducible to the Hawking temperature and life time formula for black holes in a limit $M_* = 1.531 \times 10^{-24}$ kg and $\alpha = 1$. Such that, $T = \frac{1.229 \times 10^{23} \text{kg}}{M_{BH}}$ °K and $t = 5.238 \times 10^{-19} M_{BH}^{3}$. In other words M_* is assumed to be the scale of the underlying theory. The predictions of the Hawking radiations for a black hole with mass 4.7×10^{11}kg are as given in table 2 at a mass scale 1.531×10^{-24}kg. These take on similar properties for the Higgs boson. Therefore observations at such a scale could shed more light on the detection of a 4.7×10^{11}kg black hole. If we observe at a scale of a Pion $M_* = 2.41 \times 10^{-28}$kg at the current age of the universe (about 13.8×10^9yrs) we should be able to detect a black hole with a mass of 6.698×10^4kg (Primordial Black hole).

The ideas presented above could provide a unique probe of at least four areas of physics: the early Universe; gravitational collapse; high energy physics; and quantum gravity. The first topic is relevant because studying primordial black hole formation and evaporation can impose important constraints on primordial inhomogeneities and cosmological phase transitions. The second topic relates to recent developments in the study of "critical phenomena" and the issue of whether primordial black holes are viable dark matter candidates. The third topic arises because primordial black hole evaporations could contribute to cosmic rays, whose energy distribution would then give significant information about the high energy physics involved in the final explosive phase of black hole evaporation. The fourth topic arises because it has been suggested that quantum gravity effects could appear at the TeV scale ($M_* = 1.531 \times 10^{-24}$kg) and this leads to the intriguing possibility that small black holes could be generated in accelerators experiments or cosmic ray events, with striking observational consequences (see B.J.Carr, 2005).

Lastly, the significance of α-the numerical coefficient can be seen in a broad sense when applied to the sun. If we take the sun to be a black hole with mass 1.99×10^{30} kg and a temperature at its center of $T = 1.5 \times 10^7$ K, we obtain a mass scale of $M_* = 3.717 \times 10^{-10}$ kg, which gives a life time of $\tau = 1.700 \times 10^{58}$ sec, the time that will be taken by the sun to dissipate if the temperature given at its center was 1.5×10^7 K.

The Chandrasekhar Mass Limit

A region in the universe has a potential energy of self-gravitation,

$$E_g = \frac{M_{pl}^2 c^2}{M_{BH}} \left(\frac{2M_* m_e}{m_p^2}\right) \left(\frac{6.144\pi^3}{\mu_e^2}\right) \quad (15)$$

A star will collapse to a White dwarf when the above energy is in equilibrium with the energy due to the electron degeneracy pressure of a Hydrogen atom given as,

$$E_e = m_e c^2 \quad (16)$$

Where $\mu_e = 2$ is the average molecular weight per electron, which depends upon the chemical composition of a star. Then for, $E_g = E_e$

$$M = \frac{12.288\pi^3}{\mu_e^2} \frac{M_{pl}^2 M_*}{m_p^2}$$

In a limit for $M_* = M_{pl}$ and $\alpha = 2.837 \times 10^{15}$, we obtain the Chandrasekhar mass limit for a white dwarf star as,

$$M = \frac{12.288\pi^3 M_{pl}^3}{\mu_e^2 \, m_p^2} = 1.4 M_{sun}$$

Note that such a result is only possible in the given limit but for a limit such as $M_* = 1.531 \times 10^{-24}$ kg we obtain, $M = 4.956 \times 10^{-16} M_{sun}$ which is the mass of the Primordial black hole, providing evidence for the Hawking limit for particle emission by black holes as described previously

Planck epoch

From an expression for the life time of a black hole, it is theorized that a Black hole has a mass $M_{BH} = kM_*^{1/3}$ where k is a constant. For k=1, we have a life time of 8.019×10^{-43} sec almost the Planck time-the earliest period of time in the history of the universe).

The Bekenstein-Hawking area entropy law

From the Black hole temperature we can calculate the entropy of a black hole, the total energy of a black hole with mass M and surface area A is given as,

$$E = \frac{Ac^5 M_{pl}^2 M_* m_e}{2\pi G \hbar m_p^2 M}$$

The change in entropy when a quantity of E is added to a black hole is,

$$S = \frac{E}{T}$$

Since the temperature is known (see above) on substituting we have

$$S = \frac{Ac^3 k}{4\pi G \hbar}$$

This is the Bekenstein-Hawking area entropy formula.

The Lowest Possible Energy State of a White dwarf Hydrogen Atom

In this section we prove an existence of the minimal principal quantum number which imposes a general bound on the energy level of the Hydrogen atom and the orbital radius of an electron. The results are derived from general laws not known by the entire scientific community. The section therefore provides a relationship between the micro and macro structures of the universe at a level when the atomic mass limit is in equal proportion to the Chandrasekhar mass limit.

I won't go into details of the literature of the Chandrasekhar mass limit as these have been repeatdly written and analysed in almost a million papers about the topic. But for a brief introduction into the derivation of the Chandrasekhar mass I refer the reader to Chandrasekhar 1983 Noble prize lecture (1). Almost every aspect of a white dwarf star has been studied but there is one thing which we do not know about white dwarfs in relation to the Hydrogen atom and this is encoded in Flower's original statement;

"The black-dwarf material is best likened to a single gigantic molecule in its lowest quantum state. On the Fermi-Dirac statistics, its high density can be achieved in one and only one way, in virtue of correspondingly great energy content. But this energy can no more be expended in radiation than the energy of a normal atom or molecule. The only difference between black-dwarf matter and a normal molecule is that the molecule can exist in a free state while the black-dwarf matter can only so exist under very high external pressure"

The question is, do we have an existing relationship between the mass limit of the Hydrogen atom and the White dwarf star? If the black-dwarf material is best likened to a single gigantic molecule in its lowest quantum state, what is the lowest possible energy state at which such a relationship exists?

Briefly let us propose in formula a model to support our argument; Firstly, let the potential energy of self-gravitation of a star be given by,

$$E_g = \frac{2M_{pl}^3 m_e c^2}{M_S m_{pro}^2} \left(\frac{6.144\pi^3}{\mu_e^2}\right) \qquad (17)$$

Where M_{pl} is the Planck mass $\left(\frac{\hbar c}{8\pi G}\right)^{1/2}$, m_{pro} is the Proton mass, m_e electron mass, M_S mass of star and c is the constant speed of light

Lastly, the quantized energy of an Hydrogen atom is given by,

$$E_n = \frac{m_e K_e^2 e^4}{2n^2 \hbar^2} \qquad (18)$$

Where n is the principle quantum number which indicates the energy levels in the Hydrogen atom.

By connecting the above equations we shall be able to deduce the lowest principle quantum number in the Hydrogen atom, providing one of the first relationship between the microscope and macroscopic structures of the universe.

Equating (17) to (18) we have

$$\frac{2M_{pl}^3 m_e c^2}{M_S m_{pro}^2}\left(\frac{6.144\pi^3}{\mu_e^2}\right) = \frac{m_e K_e^2 e^4}{2n^2 \hbar^2}$$

On arranging and canceling like terms we have;

$$n = \frac{\alpha_e \mu_e m_{pro}}{2}\sqrt{\frac{M_S}{6.144\pi^3 M_{pl}^3}}$$

Where $\alpha_e = \frac{K_e e^2}{\hbar c} = \frac{1}{137}$ is the fine is is structure constant.

When, $M_s = 1.4 M_{sun}$ (Chandrasekhar mass limit), we have $n = 5.1586 \times 10^{-3}$. This is the allowed principal quantum number or the lowest energy state of an Hydrogen atom for a white dwarf star at the Chandrasekhar mass limit. Therefore the quantized energy of the Hydrogen atom at this principal number is

$$E_n = \frac{13.606 eV}{n^2} = 511.289 \times 10^3 eV$$

And the electron radius at this energy level is $r = 1.405 \times 10^{-15} m$.

This result implies that, whereas the Bohr's orbital quantization doesn't permit orbits below the Bohr radius of $5.28 \times 10^{-11} m$, the theory above says that this is possible for an atom under high pressure. The electrons are therefore bound to the surface of the proton. For a white dwarf star of $M_s = 0.87 M_{sun}$, we have $n = 4.07668 \times 10^{-3}$. This gives the radius of a proton of $r = 8.775 \times 10^{-16} m$ which has been determined by spectroscopy methods.

18. The Spectrum of the Atomic Universe

Summary

Quantum mechanics or quantum theory is a physical science that is concerned with the behaviors of matter and energy at a scale of atoms and subatomic particles/waves. Quantum mechanics also acts as the basis through which we can study, analyze and explain very large objects such as stars and galaxies, and cosmological events such as the big bang. To describe the atomic universe fully we need both quantum mechanics and gravity. This is achieved through the study of the accelerations of the particles leading to the radiations of electromagnetic energy, and predicting that all matter is unstable. It is then theorized that there appears two accelerations, g_a and g_b whose ratio explains the formation of two or more lines close together in the Hydrogen spectrum which is known as the fine structure. Comparing this model with Bohr's model of the Hydrogen atom produces very precise results for cosmological events hence the atomic universe.

Introduction

In the early 20th century, Ernest Rutherford experiments established that atoms consisted of a diffuse cloud of negatively charged electrons surrounding a small, dense, positively charged nucleus. Given this experimental data, it was quite natural for Rutherford to consider a planetary model for the atom, the Rutherford model of 1911, with electrons orbiting a sun-like nucleus. This model was a difficulty. The laws of classical mechanics predict that the electron will release electromagnetic radiation as it orbits a nucleus. Because the electron would be losing energy, it would gradually spiral inwards and collapse into the nucleus. This is a disaster, because it predicts that all matter is unstable.

To overcome this difficulty, Niels Bohr proposed, in 1913, what is now called the Bohr model of the atom. The model's key success lay in explaining the Rydberg formula for the spectral emission lines of atomic hydrogen. Not only did the Bohr model explain the reason for the structure of the Rydberg formula, but it provided a justification for its empirical results in terms of fundamental physical constants.

This paper looks at the model in a very different way than that of Bohr. The fact that all accelerated particles do emit electromagnetic radiations is taken into account and therefore the acceptance for the unstableness of all matter is considered in due respect. In fact Bohr's ideas never required classical mechanics simply because it could not conform to the experimental observations of the spectrum of the Hydrogen atom that were obtained by Rydberg using his formula.

To merge gravity with Planck's quantum theory by then was also a problem at hand and therefore Bohr had to forego the problem by introducing in his theory adhoc postulates, and this could have been the reason why Einstein found problems in merging gravity with electromagnetism in what is called "The Grand unified field theory", of which he had to question the problem with the quantum theory and therefore request for a complete quantum theory. From Bohr's model many theories have been formed each building from the ideas of the model, but a certain point is reached where the theories can not conform well to the known laws of nature and therefore regarded as failures, which of course in their judgments is true. The problem is seen to come from exactly the roots of quantum mechanics.

The aim of this paper is therefore to produce a generalized theory of atomic structure that incorporates in it gravity and quantum mechanics. In other words a theory that takes the laws of classical mechanics into consideration.

Methodology

The Hydrogen atom exists in certain stationary states of discrete energies. The acceleration due to gravity of an electron in orbit around the nucleus will cause the atom to emit radiations (radiate energy) and thus make the atom unstable. The acceleration (g) falls off with time t provided the radius of orbit of the electron R is a constant thus the acceleration due to gravity is given by;

$$g = R/\Delta t^2 \qquad (1)$$

The rate of change of energy P radiated as a result of the above acceleration will depend on the constants c (speed of light) and G (universal gravitational constant), hence;

$$P = c^5/G \qquad (2)$$

The power and time must be re-quantized in units of $\hbar = h/2\pi$ where h is Planck constant, hence

$$P\Delta t^2 = n^2 \hbar \qquad (3)$$

Where n= 1,2,3…….. is the principle quantum number.

But the total energy of the atom in the various energy states is $W= -ke^2/R$ where k is the Coulomb constant and e is the elementary charge. Since Δt^2 is known from Eqn1 and P from Eqn2 then using Eqn3 the radius is given by

$$R=n^2Gg\hbar/c^5 \qquad (4)$$

From which the total energy is given by,

$$W=-ke^2c^5/n^2Gg\hbar \qquad (5)$$

From the Bohr-Einstein frequency (f) condition, applied to a transition from a level with $n = n_i$ to a level with $n = n_f$, The energy of a photon emitted by a hydrogen atom is given by the difference of two hydrogen energy levels

$$hf = E_i - E_f$$

Finally we get since frequency $f = c/\lambda$, where λ is the wavelength

$$1/\lambda = [ke^2c^4/2\pi G\hbar^2][1/g][1/n_f^2 - 1/n_i^2] \quad (6)$$

The equation obtained above shows some how a great significance of gravity in the quantum theory. So far it states that regardless of the levels in the transitions of an atom the acceleration due to gravity of the particles in the atom do greatly affect the nature of its spectrum.

Results

The quantity $[ke^2c^4/2\pi G\hbar^2]$ is the inverse of the square of time t and therefore

$1/t^2 = [ke^2c^4/2\pi G\hbar^2]$, from which the time is obtained as $t = 1.58873 \times 10^{-42}$s.

Comparing Eq6 with Bohr's model, here we shall equate the Rydberg constant $[k^2e^4m/4\pi c\hbar^3]$, where m is the mass of the particle, to the constant $[ke^2c^4/ 2\pi G\hbar^2][1/g]$. Doing this generates an acceleration given by $g_a = [2\hbar c^5/ke^2 Gm]$ from which we obtain a general equation of forces given by $[8\pi G/ c^4][gm][ke^2/R^2] = 16\pi\hbar c/R^2$. where $[gm]$ is the gravitational force and $[ke^2/R^2]$ is the electromagnetic force.

At the Schwarz child's radius $R=Gm/c^2$ the acceleration is $g_b = c^4/Gm$ which gives an equation for the spectrum as $1/\lambda = [/ 2\pi a_o][1/ n_f^2 - 1/ n_i^2]$ where a_o is the first Bohr radius $[\hbar^2/ mke^2] = 5.28 \times 10^{-11}$m.

The interesting part of it is that the ratio $g_b / g_a = [ke^2/2\hbar c]$ the fine structure constant. This result therefore explains the fine structure shown by the Hydrogen spectrum and thus suggests that an electron describes an elliptical orbit. Now using the acceleration g_a, the radius a_o and the mass m the energy W of a particle will be given by $W = g_a a_o m = \beta/m$, where β is a constant given by $[2\hbar^3c^5/k^2e^4G] = 1.64367 \times 10^6$Jkg. For two different masses m and m_o we have the equation for the product of the masses as, $mm_o = [2\hbar^3c^3/k^2c^4G]$.

Discussion

The results obtained give out a clear image for the description of the atomicity of both large and small particles. Firstly the time t obtained is the is the earliest period of time in the history of the universe from zero to approximately 10^{-43} seconds, during which quantum effects of gravity were significant. At this period all the fundamental forces of physics were unified. The state of the universe during this epoch was unstable, tending to evolve and giving rise to the familiar manifestations of the fundamental forces through a process known as breaking. Symmetry breaking quickly led to the era of cosmic inflation, the Inflationary epoch, during which the universe greatly expanded in scale over a very short period of time.

Secondly, the accelerations g_a and g_b led to different spectrums of the Hydrogen atom. Where g_a produces the Rydberg equation for the spectrum of the hydrogen atom, that is incorporated in it the Rydberg constant [$k^2 e^4 m/4\pi c \hbar^3$], the acceleration g_b produces a different equation which instead of a Rydberg constant, it has the inverse of the first Bohr radius a_o. These differences in the spectrum of the hydrogen atom with the former producing a single line and the latter two or more lines of the spectrum close together, imply that the electron moves in an elliptical orbit as those of the planets in orbit around the sun, hence the ratio of the accelerations g_b / g_a will generate a fine structure constant describing the closeness of the spectrum lines produced by the hydrogen atom. Finally the energy obtained using the acceleration g_a and the first Bohr radius a_o has an impact on the way we express the energy of large and small particles. For example a body of one kilogram mass (1Kg), will have an energy of 1.64367 ×10^{6}Joules (1.64367 ×10^{6}J) which is a very high energy. This energy is independent of the speed of a body or particle in question, and thus gives the energy to a particle regardless of it's speed. We very well know that the speed of light is a constant and therefore doesn't change and that with relativity such a body of 1kg will have energy approximately 10^{16}J which is lager than the first one by 10^{10}. For smaller particles say an electron we have from the result equation the energy as 1.80425×10^{36}J, but for relativity it is $\sim10^{-15}$J. These differences in energies imply that without knowing the speed of the particle we can obtain it's energy depending on it's mass since some particles tend to move at a speed greater than that of light.

Conclusion

In conclusion the results produced successfully indicate that without gravity, quantum mechanics cannot survive and without quantum mechanics, gravity cannot survive. Therefore the two theories are needed to explain the atomic universe fully.

19. On the Generalization of Loop Quantum Gravity

Quantum gravity is the field of theoretical physics that tries to unify quantum mechanics with general relativity. Quantum mechanics describes the three fundamental forces of nature while general relativity is a theory of the fourth fundamental force: gravity. The goal everyone is waiting for to emerge from this unification is a "theory of everything", or "Grand Unified Theory" (GUT). In 1986, Abhay Ashtekar reformulated Einstein's field equations of general relativity using what have come to be known as Ashtekar variables, a particular flavor of Einstein-Cartan theory with a complex connection. He was able to quantize gravity using gauge field theory. In the Ashtekar formulation, the fundamental objects are a rule for parallel transport and a coordinate frame known as a vierbein at each point. Because the Ashtekar formulation was background-independent, it was possible to use Wilson loops as the basis for a nonperturbative quantization of gravity. Explicit (spatial) diffeomorphism invariance of the vacuum state plays an essential role in the regularization of the Wilson loop states.

Around 1990, Carlo Rovelli and Lee Smolin obtained an explicit basis of states of quantum geometry, which turned out to be labelled by Penrose's spin networks. In this context, spin networks arose as a generalization of Wilson loops necessary to deal with mutually intersecting loops. Mathematically, spin networks are related to group representation theory and can be used to construct knot invariants such as the Jones Polynomial.

The need for this chapter is to understand those problems involving the combination of very large mass or energy and very small dimensions of space, such as the behavior of black holes, and the origin of the universe.

The formula for the quantization of quantum gravity

The model is based on separating the gravitational field into the sum of two components; that is the background and the quantum field. The background left is one for all our calculations. But because loop gravity ignores the back ground space as a lost entity that does not occur in space, there fore the need to reconstruct quantum field theory from scratch without a background space is taken into account. I therefore suggest that the calculation should be performed by summing all possible space-times.

Quantum field theory depends on particle fields embedded in the flat space-time of special relativity. General relativity models gravity as a curvature within space-time that changes as a gravitational mass (m) moves. Assuming a spherical symmetric object that space time is of dimensions increasing from 1, 2, 3, 4...N, where N is the nth term of the dimensions. To quantize space and time is to create a space in which all of physics is quantized. The nature of the curved space surface is described by increasing powers in the Schwarzschild radius $R_s = Gm/c^2$, Hence describing the dimensions of space. Quantum mechanics explains the existence of discrete energy states in an atom, in away that the angular momentum of the atom must be quantized, which is also the case for quantum gravity. The equation for the quantization of the loop quantum gravity can then be written as,

$$\eta R_s + \beta R_s^2 + \mu R_s^4 + \ldots\ldots\ldots + \delta R_s^N = n\hbar \qquad [127]$$

Where $\eta = \sqrt{Beh}$, is the momentum of a particle probing another form of quantum mechanics, $\hbar = h/2\pi$, where h is Planck constant, $\beta = 8\pi Be$, e is the elementary charge, B is the magnetic field and finally $\mu = 256\pi^3 P/c^2$, where P is the intensity and c is the constant speed of light.

The energy equation

What changes is the form of the equation the rest remaining constant. The principle behind this is that eqn1 can be changed to any form simply for purposes of calculating complex phenomenon. The energy to which we are concerned here is expressed as a general expression describing the energy scales forming smaller and larger matter entities in the universe. The energy will thus be given by;

$$\eta c + \beta c R_s + \mu c R_s^3 + \ldots\ldots\ldots + \delta c R_s^{N-1} = n\hbar c / R_s$$
[128]

Note: the background space described by the Schwarzschild radius has changed, thus the above equation in any case can be used to calculate the basic properties of Black holes. Remember the Schwarzschild radius is the radius for a given mass where, if that mass could be compressed to fit within that radius, no known force or degeneracy pressure could stop it from continuing to collapse into a gravitational singularity.

The mass equation

Having explored the energy scale we now form general equation that describes well the mass scale. This is also done the same way as eqn128 and therefore generate,

$$\eta/c + \beta R_s/c + \mu R_s^3/c + \ldots\ldots + \delta R_s^{N-1}/c = n\hbar/cR_s \qquad [129]$$

The maximal magnetic field

Assuming that the energy $W = \beta c R_s$, from eqn128 is equal to the energy $W = mc^2$, we hence obtain the magnetic field as, $B = c^3/8\pi Ge = 1.0054 \times 10^{53}$ N/Am. using this magnetic field in the energy equation, $W = \eta c$ we get the energy in the form $W = (c^2/2)\sqrt{\hbar c/G}$ where the quantity $\sqrt{\hbar c/G}$ is the Planck mass M_p at an energy of 6.119×10^{18} Ge

Time taken by a black hole to evaporate and its entropy

The energy required here is given in Eqn128, it is at this, that the intensity $P = W/A\Delta t$, (where A is the area and t is the time) is used. We take the energy $W = \mu c R_s^3$ (from Eq128) as our interest from which we obtain the time as $\Delta t = 256\pi^3 R_s^3/Ac$. But with black holes the area will become exactly equal to the square of the Planck length as $A \sim L_p^2 = \hbar G/8\pi c^3$ hence the change in time is given by $\Delta t = 63500.86\pi G^3 m^3/\hbar c^4$.

For entropy we set the energy to kT, where k is Stefan's-Boltzmann's constant and T is the temperature of the body. Now for $kT = \mu c R_s^3$, since Δt is known the entropy is thus given by $S = W/T = 78.96 Ak c^3/ \pi \hbar G \sim A/4$. In conclusion we state that the entropy of a black hole is proportional to the area of the event horizon.

The quantum Hall Effect

For this effect the momentum η is used. From Eqn128 we set, $\eta c = n\hbar / R_s$ which gives the magnetic flux as $4 \pi R_s^2 B = nh/e$, from which the resistance is given by $\zeta = 4 \pi R_s^2 B /e = nh/e^2$. for n= 1,2,3,4 the resistance is of a value 25833.8Ω

Maximum Intensity

Using eqn129 in this case, since B is known and P got from $\mu R_s^4 = n\hbar$; as $P = \hbar c^2/256\pi^3 R_s^4$, we hence obtain, $M_p /2 + m + M_p /m = M_p /m$, which gives $M_p + 2m = 0$, and for identical mass M =0, which is true. The intensity at the planck length that is for $R_s = L_p$ is $P=c^8/\pi \hbar G^2$

20. An Exceptionally Simple Classical Unified Field Theory

Introduction

The development of a unified field theory of physics over so many years has involved the use of complicated mathematics coupled with assumptions and imaginations that have made it impossible for the creation of a simple and beautiful unified field equation. The mathematics for unification should be concise and elegant.

Attempts to unify gravitation and electromagnetism came a long way with the realm of Albert Einstein in what was called a classical unified field theory. However such attempts were abandoned with the development of modern theories of physics i.e. The General Relativity theory and the Quantum theory. These two theories are elegant and predictive in their domain of applicability.

A classical unification theory should reproduce these two theories and also be able to explain it. Today we know of the standard model and general relativity and how they are successful, but the unification of the two theories has also proved to be difficult. In my conscience these should be fruits of a classical unified field theory. In other words a classical unification incorporates them as one that is they are supposed to appear in the unification equation. Therefore the development of modern physics acts as a check for classical physics in any field of physics concerned with unification.

The objective of this paper is to develop a unified field theory uniting gravity and electromagnetism using classical physics and non- quantum approaches.

The unified electromagnetic gravitational oscillations

The gravitational potential field (gravity) is caused by the presence of any mass in space. Like gravity, charge is a fundamental property of the universe. Just as mass causes a gravity field, charge causes an electromagnetic field. There are no known methods of insulation of gravitational field in modern science. It is impossible to imagine space and gravitation separately. Gravitation exists everywhere where there is some space. The gravitational field created by all masses of our metagalaxy is the aether in which cosmic objects moves and electromagnetic oscillations are propagated. The space surrounds us since the whole matter carries gravitational charge of only one sign.

A dipole antenna is used to produce overlapping electric and magnetic fields E and B respectively. Such an antenna is so far the simplest practical antenna from a theoretical point of view, and therefore is the basis on which this study is conducted. The current amplitude I for such an antenna decreases uniformly from maximum at the center to zero at the ends.

The work done to move the electron at a speed v(t) up and down the antenna is the gravitational potential energy W_G and is proportional to the velocity v(t). The electric potential energy W_E of the electron at a distance R from the antenna is proportional to their angular momentum L. $W_G = k_1 v(t)$ and $W_E = k_2 L$, if the proportionalities are true then also the following should be true $W_G W_E = \beta L(t) v(t)$ where $k_1 k_2 = K^2 = \beta$ is the force required to accelerate the electrons.

Differentiating W_G with respect to height h moved by a charge up the antenna, and W_E with respect to the charge q on the electron i.e. $dW_G/dh = mg$, $dW_E/dq = ER$. Then the product of the acceleration due to gravity g and the electric field E is

Eg = (βL/mRt)div   1

∇ = div = d/dq

When a charge is moved up and down the antenna by the oscillator in the y direction a changing magnetic field is produced in the z direction leading to a changing gravitational field. Then the acceleration due to gravity is of magnitude g= vcΩ, Ω is the curvature of space in regions near the antenna where a charge experiences the electric and magnetic field. And c is the speed of the electromagnetic wave transmitted from the antenna. Substituting for g in equation 1 gives

$\nabla \beta = d\beta/dr = f_e \Phi$ 2

This is the rate of change of force with the radius of a spherical wave front r=cT, T is the time- dimension and then $r^2 = x^2 + y^2 + z^2 + T^2$. Therefore $\nabla \beta$ depends on the electromagnetic force f_e=Eq and the curvature $\Phi = \sqrt{BI/W}$. where W is the kinetic energy of the electron and I is the current.

According to R.L. Collins, (1997-2006) Energy is exchanged between neutral masses, via a long range electromagnetic force, and that this exchange of energy reproduces the effects of gravity. If β is taken as the gravitational force acting perpendicular through the center of mass of the particle then the potential gravitational gradient will be given by

$$\nabla^2 W_G = f_e \Phi \qquad \qquad 3$$

The electromagnetic wave transmitted from the antenna will move along the x-axis. Since this wave is moving through a vaccum then there can be no conventional currents, but there is a displacement current, hence I =0 and Φ=0. Then the condition $\nabla^2 W_G$ =0 is fulfilled. The accelerated motion of electrons up and down a straight-rod dipole antenna therefore produces electromagnetic waves that satisfy the above equation. The electric, magnetic and gravitational fields are periodic in both time and space.

Discussion and conclusions

Change of both electric and gravitational field results in the creation of a magnetic field in the region of space-time which has a dual electrogravitational nature. The amplitude of electromagnetic and magnetogravitational constituents of the unified electromagnetic gravitational oscillations depends on field potential of opposite nature. The electromagnetic constituent is determined by gravitational potential and the magnetogravitational one is determined by electric potential. Transference of gravitational masses of matter in electrogravitational field-aether causes the creation of the proper magnetic field. (V.Ya.Kosyev, 2000).In the quantum realm, the gravitational force is so weak that it is difficult to observe quantum effects caused by gravity. However Nesvizhevsky and collaborators have reported an experiment in which they observed quantum effects of gravity on the behaviour of ultracold neutrons (UCNs). These neutrons have kinetic energies so low that they can be trapped by gravity above a reflecting surface. (Thomas J. Bowles, 2002)

21. The earliest period of time in the history of the Universe

A simple mathematical model is used to describe the relationship between the forces at the Planck epoch and the grand unification epoch. This model explains the problems concerned with what the accelerations and wave length of the particles would have been in the early universe. The forces are found to depend on the time at any given scale in the epoch and when these times are equal the fine structure constant is generated.

Introduction

This section represents a brief history of the universe, that is from its past to present. Observations have suggested that the universe began 13.7billion years ago. The universe was so hot with particles having a very high energy, in its earlier phase. The evolution then proceeded with this energy forming the first protons, electrons and neutrons, then nuclei and finally atoms. The microwave background was also emitted during the formation of the neutral hydrogen. Finally the structure of the universe was formed when matters aggregated into the first stars and quasars and on large scale clusters of galaxies and super clusters were formed.

In cosmology, the Planck epoch , named after Max Planck, is the earliest period of time in the history of the universe, from zero to approximately 10^{-43} seconds, it is at this time that quantum effects of gravity were significant. At this period approximately 1.37×10^{10} years ago all fundamental forces were unified. The state of the universe during the Planck epoch was unstable, tending to evolve and giving rise to the familiar manifestations of the fundamental forces through a process known as symmetry breaking. It is currently believed that the Planck epoch inaugurated the Grand unification epoch, and that symmetry breaking quickly led to the era of cosmic inflation, the Inflationary epoch, during which the universe greatly expanded in scale over a very short period of time.

The age of the universe, in Big Bang cosmology, refers to the time elapsed between the Big Bang and the present day. Current observations suggest that this is about 13.7 billion years, with an uncertainty of about +/-200 million years. Extrapolation of the expansion of the universe backwards in time using general relativity yields an infinite density and temperature at a finite time in the past. This singularity signals the breakdown of general relativity. How closely we can extrapolate towards the singularity is debated—certainly not earlier than the Planck epoch. The early hot, dense phase is itself referred to as "the Big Bang", and is considered the "birth" of our universe. Based on measurements of the expansion using Type Ia supernovae, measurements of temperature fluctuations in the cosmic microwave background, and measurements of the correlation function of galaxies, the universe has a calculated age of 13.7 ± 0.2 billion years.[21] The agreement of these three independent measurements strongly supports the ΛCDM model that describes in detail the contents of the universe.

The aim of this section is to examine the Planck epoch and the grand unification epoch and therefore find out the true scale that explains the earliest period of the universe.

Methodology

I construct a mathematical model to study the relationship between the ratio of the wavelength and the ratio of the accelerations. In the model I refer to different particles probing different scales of time, I refer to the Planck scale and the quantum gravity scale. Assuming two particles, one under the influence of gravity and the other the influence of quantum fields, when the two particles are set to fall through the fields each falls with an acceleration describing the fields that is g_Q and g_G respectively. The length scales describing the falls are λ_e and λ_c, so the equation below will fully describe the model,

$$\lambda_e/\lambda_c = (g_Q/g_G)^{1/2} \qquad [1]$$

where, $\lambda_e = e/2\lambda_p E\varepsilon$, is the wavelength of an electron depending on the wave length of a photon λ_p. The higher the photons wavelength the smaller the electrons wavelength and the smaller the wavelength of the photon the higher that of an electron, this takes place on the assumption that the electric field

E, the charge e, and the permittivity of free space ε, are all constants. $\lambda_c =$ ℏ/mc, is the Compton wavelength of an electron with mass m , here ℏ is Dirac constant and c is a constant speed of light. g_Q = $e^2 f/2$ ℏε describes the acceleration of a particle with a frequency f in the quantum field and , g_G = Gm/R^2 is the acceleration due to gravity and G is the universal gravitational constant.

for $\lambda_p = c/f = 2\pi R$, $E = e/4\pi\varepsilon R^2$ where R is the distance between any two particles.

We compare the forces that come as a result of the motion. There are two forces each occurring at a different scale length. First there is a force describing the Inflation, baryogenesis at the grand unification transition. Second there is another force describing the quantum gravity barrier at the Planck epoch. Both of these forces will have a similar characteristic, which is they will depend on the time t at any level.

Therefore from eqn1 we multiply through by Gc^5 and obtain a relationship of the forces given by,

$$F = t_g(c^7/32\pi mG^2) = t_u(c^7/16\pi G^2 m) \quad [2]$$

Where

$t_u = e^2\hbar/2\varepsilon Gm^3 c^2$ and $t_g = A/cR$, A is the area.

Results

At Planck scale mass $m_p = \sqrt{\hbar c/G}$, $t_u = (e^2/\varepsilon)\sqrt{G/\hbar c^7} = 4.932\times 10^{-45}$s and at the Planck length scale $L_p = R = (\sqrt{\hbar G/8\pi c^3})$, $A \sim L_p^2 = \hbar G/8\pi c^3$ and $t_g = \sqrt{\hbar G/c^5}$, then equating t_u to t_g the fine structure constant is generated as $(e^2/16\pi^2 \varepsilon \hbar c)$.

Discussion

The results obtained show exactly the required mechanism responsible for the description of the time line of the big bang. The reduction of the fine structure constant from the theory shows that there exists a relationship between quantum mechanics and gravity, that at the two times t_u and t_g the force of gravity was strong and that there was a possibility for the unification of all the fundamental forces of nature. It can now be theorized that there exists a scale that when merged with the Planck scale the result is the earliest period of the universe at which all of physics problems can be solved. Therefore both scales are needed to explain the origin of the universe from the big bang.

Conclusion

Ignoring quantum effects means that the universe starts from a singularity with an infinite density. This hypothesis however can change when quantum gravity is taken into account. The works of String theory , Loop quantum gravity, Noncom mutative geometry and other fields of physics holds promise for our understanding of the very beginning.

However; the more we understand about how matter forms, the more precisely we will be able to interpret what we learn from astrophysical data, and from other sources.

22. Another form of general relativity and its new predictions

This study addresses the issue of analyzing galactic features and the big bang whose portfolio is related to the age and density of the universe. The age of the universe and its features that is density, pressure, temperature and its epoch are all formulated from first, considering the curved nature of space and time. And second modifying the first law of thermodynamics to include the features describing the large objects and small particles, It is then seen that the area occupied by heavenly bodies in space divided by a dimensionless constant determines the tidal force acting on that body and that Charles's and Boyles law of an ideal gas when applied to black holes gives its entropy with a gravitational coupling constant determining the strength of the gravitational field.

Introduction

The development of general relativity followed a publication of acceleration under special relativity in 1907 by Albert Einstein. In his article he argued that any mass will "Distort" the region of space around it so that all freely moving objects will follow the same curved paths curving toward the mass producing the distortions. The questions raised by the principle of equivalence and general relativity are intimately related to the questions of the origin, size, and structure of the universe. Is the universe infinite or finite? How old is our solar system and galaxy? How were they formed? How many other galaxies are there and how are they distributed? Where did they come from? What was the universe like before these galaxies were formed? The field of physics that deals with these questions is called cosmology, a very fast moving field.

In 1916, Schwarzschild found a solution to the Einstein field equations, laying the groundwork for the description of gravitational collapse and, eventually, black holes. In 1917, Einstein tried to describe a static universe, where he added cosmological constant to his original field equations for that purpose. With Hubble's observations in

1929, on the movement of galaxies which predicted an expanding universe, Lemaître formulated the earliest version of the big bang models.

General relativity uses a complex mathematical equation that makes it so hard for people to master the theory. This paper gives out a simple and accurate mathematical formulation of space and time is some what a similar fashion to that of general relativity. This paper deviates from the theory in that for it takes into account the description of space and time for both small (quantum effects) and large particles. This paper also explains the features of cosmology (black holes) and the Big bang (the earliest period of the universe).

Finally, there have been various attempts through the years to find modifications to general relativity. The most famous of these are the Brans-Dicke theory and Rosen's bimetric theory. Both of these proposed changes to the field equations, and both suffer from these changes permitting the presence of bipolar gravitational radiation. As a result, Rosen's original theory has been refuted by observations of binary pulsars. As for Brans-Dicke the amount by which it can differ from general relativity has been severely constrained by these observations. It is generally held that one of the most important unsolved problems in modern physics is the problem of obtaining the true quantum theory of gravitation, that is, the theory chosen by nature, one that will work at all energies. Discarded attempts at obtaining such theories include supergravity, a field theory which unifies general relativity with supersymmetry. In the second superstring revolution, supergravity has come back into fashion, with its as yet undefined quantum completion rebranded with a new name: M-theory.

Materials and methods

The movement of a particle in a curved path and their associated forces

Since gravity increases in inverse proportion to volume, any quantity of matter that is sufficiently compressed will become a black hole. When a large enough amount of mass is present within a sufficiently small region of space, all paths through space are warped inwards towards the center of the volume, forcing all matter and radiation to fall inward. I formulated a new solution to Einstein field equation which describes black holes, and is given by;

$$\text{Volume} = A_B R_d = (1/F_e)(h^2/m) \quad [1]$$

Where A_B is the area of the small region of space, F_e is the tidal force (An object in any very strong gravitational field feels a tidal force stretching it in the direction of the object generating the gravitational field.) Near black holes, the tidal force is expected to be strong enough to deform any object falling into it, even atoms or composite nucleons; this is called spaghettification. The strength of the tidal force depends on how gravitational attraction changes with distance, rather than on the absolute force being felt. This means that small black holes cause spaghettification while infalling objects are still outside their event horizons, whereas objects falling into large, supermassive black holes may not be deformed or otherwise feel excessively large forces before passing the event horizon.

$R_d = A_e^2 / R_s^3$ is the radius of that region of space, $A_e = hc/F_e$ is the area occupied by each particle experiencing the tidal force (in other words area of the object). $R_s = Gm/c^2$ is Schwarzschild radius, It is the radius for a given mass where, if that mass could be compressed to fit within that radius, no known force or degeneracy pressure could stop it from continuing to collapse into a gravitational singularity, h is the Planck constant, m is the mass of the object, c is the speed of light and G is the gravitational constant. It should be noted that as the volume R_s^3 increases the radius R_d reduces and as it reduces the radius increases. Therefore R_d is the radius of a region of space that is changed when ever the volume occupied by a compressed mass in that region changes.

Therefore the area is given by,

$$A_B = (R_s^3/A_e^2 F_e)(h^2/m) = G^3 m^2 F_e/c^8$$
[2]

The equation obtained shows how the force depends on the area where the mass is concentrated. The above force differs from Newton's gravitational law in that it is directly proportional to the area but inversely proportional to the square of the mass of the body. Hence $F_e = A_B c^8 / G^3 m^2 = N A_B / m^2$ where $N = c^8 / G^2$

Results

The area gives the forces

Since in Eq2 the force is related to the area we can then use it to obtain the force on any object occupying any given area. If we take a square of the Schwarzschild radius to be the area where if a mass could be compressed to fit within that area, no known force or degeneracy pressure could stop it from continuing to collapse into a gravitational singularity, then the following is obtained

For a black hole of area R_s^2, the force is $F_e = c^4 / G$

For two particles separated by a distance R and within an area R_s^4 / r^2, Newton's gravitational force is $F_e = Gm^2 / r^2$

For particles probing the big bang, the areas are R_s^2/α_s and R_s^2/α_g (where $\alpha_g = Gm^2/\hbar c$ is the gravitational coupling constant and $\alpha_s = ke^2/\hbar c$ is the fine structure constant, k is coulomb constant and e is elementary charge) the following forces are obtained respectively,

$$F_G = E_o^2/ke^2 \text{ and } F_E = E_o^2/Gm^2$$

the energy $E_o = \sqrt{\hbar c^5/G}$ where $\hbar$ is the Dirac constant $h/2\pi$. This is the energy describing the scale of the energy that the universe had in its early formation. Therefore substituting the value of F_E in Eq2 we obtain $A_{B1} = \hbar G/c^3 = 2.60624 \times 10^{-70} m^2$ And for $F_G = F_e = E_o^2/ke^2$ we obtain $A_{B2} = A_{B1}(Gm^2/ke^2)$. This means that the force F_E only becomes comparable to F_e at the Planck length scale. And $F_G = F_e$ doesn't achieve the correct scale when the forces are compared, it approaches the scale but with an effect Gm^2/ke^2 which indicates that the gravitational force cannot be compared to the electromagnetic force.

This therefore states that F_G is the gravitational force and F_E is the electromagnetic force at the big bang scale.

The cosmological pressure and temperature

From Eq2 the pressure P is simply a tidal force on an object per unit area occupied by the object in a region of space, hence,

$$P \quad = \quad F_e/A_B = \quad c^8/G^3m^2 \quad [3]$$

And finally the temperature from Eq1 is

$$T \quad = \quad (R_d F_e/k) \quad = \quad (1/A_B)(h^2/mk) \quad [4]$$

Where k is gas constant per molecule in joules per Kelvin

The entropy of a black hole and the first law of thermodynamics

Entropy of a black hole

Keeping the volume constant, the pressure of a gas is directly proportional to its absolute temperature that is $P \propto T$, hence from Eq3 and Eq4

$$P/T = (\beta^2 A_B/R_s)(k/h^2) \quad [5]$$

Where β is the rate of change of mass equal to c^3/G

For an ideal gas, keeping the temperature constant, the volume of a gas will vary inversely proportional to its pressure that is V α 1/P, hence

$$PR_s^3/T = (\beta^2 A_B R_s^2)(k/h^2) \quad [6]$$

The above equation is the entropy of a black hole derived from the properties of a gas and therefore it can be expressed as

$$\text{Entropy} = PR_s^3/T = (kA_B/4A_{B1})(2\beta R_s^2/\pi h) = (kA_B/4A_{B1})\alpha_g \quad [7]$$

Where $\alpha_g = (2\beta R_s^2/\pi h)$ is the gravitational coupling constant.

The first law of thermodynamics

The sum of the kinetic energy and potential energy of all the individual particles making up the system is the internal energy given by

$U = \Delta Q + \Delta W$

Where ΔQ is the heat flow into the system and ΔW is the work done by the system. Basing on the results obtained

$\Delta Q = \sqrt{\hbar c^5/G} = c\sqrt{\beta \hbar} = E_o$ and $\Delta W = -(\beta^2 A_B R^2_s)(kT/h^2) = -P R^3_s$.

Hence the internal energy is formulated as

$U = (\beta \hbar / E_o)(c^2 - R^2_s E_o [kT/h^2 \quad])$ [8]

Letting $[kT/h^2] = 4\pi^2 / \hbar \tau$ where τ is the time

We obtain

$$U = (\beta\hbar/E_o)(c^2 - R_s^2 E_o[4\pi^2/\hbar\tau]) \quad [9]$$

as a result when $U = 0$, $R_s = 1.61414 \times 10^{-35}$ m, and $E_o = 1.9605 \times 10^9$ J. the time

$\tau = 2.1238 \times 10^{-42}$ s, which is the earliest period of the universe is obtained.

Still from Eq8 we find that the quantity $(\beta\hbar/E_o)$ represents mass which is given by $M_p = 2.1765 \times 10^{-8}$ kg. Multiplying this mass throughout we generate a principle equation

$$U = M_p c^2 - M_p (R_s^2 [E_o kT/h^2]) \quad [10]$$

This equation gives us a mechanism of combining the laws governing small particles (quantum mechanics) with those governing heavenly bodies (General relativity). The appearance of the Schwarzschild radius R_s which explains galactic bodies, the appearance of the random energy kT that describes small particles, the appearance of the Planck mass M_p and energy E_o which describe the Planck epoch in the early universe are all evidence of the generalized formulation of the combined theory of quantum and gravity hence obtaining a quantum gravity theory of nature.

From which we get the speed of particles in the early universe given by

$$\upsilon = \sqrt{(R_s^2 [E_o kT / h^2])}$$

For $E_o = kT$ the speed is got as $\upsilon = R_s E_o / h = 0.4773 \times 10^8$ m/s which is smaller than the speed of light by only ($\upsilon/c = 0.1591$)

Discussion

From the results obtained it is studied that the forces acting on heavenly objects depend on the areas of space in which these objects occupy. These areas are also related to the "Schwarzschild area" R_s^2. any area in space will have this effect and it will only change when R_s^2 is divided by a dimensionless constants, for example, the Newtonian gravity will depend on the dimensionless constant R_s^2/r^2, the cosmological features depend on a unity dimensionless constant and For forces describing the big bang the dimensionless constants will be the coupling constants determining the strength of the gravitational and electromagnetic fields. When all these dimensionless constants are equal to unity it means that the area occupied by one object in the universe corresponds directly to that occupied by other objects and that the effect of the force to one object is the same to all other objects, therefore implying that the forces will then be unified into one fundamental force.

Forces probing the big bang are directly related to the square of the Planck energy, and it is known that at the Planck scale the description of subatomic particle interactions in terms of quantum field theory breaks down, but since both forces have energies at the Planck scale it means that the two are comparable to the other forces and when they are equated the result is a dimensionless constant which is unity $Gm^2/ke^2 =1$.

The thermodynamic laws that describe gases here on earth are seen to be the same laws that govern the particles found in our galaxy. It is seen that Boyle's and Charles laws can be applied to heavenly bodies, the result of this is that the entropy of these bodies is directly proportional to the product of the area of the event horizon of the body and the gravitational constant that determines the strength of the gravitational force.

The earliest period or time line of the big bang is studied. The random energy of particles kT forming matter during that time was in equilibrium with the energy of the photons, the time when this happened was 2.1238×10^{-42}s, every particle during this time moved at a speed of light $c=3\times 10^8$ m/s, particles moving at a speed closer to that of light where produced when kT was equal to $E_o = 1.9605 \times 10^9$ J and when calculated had a speed $\upsilon = R_s E_o / h = 0.4773 \times 10^8$ m/s which varies directly with the Schwarzschild radius.

Conclusion

Successfully I have analyzed a method of combining elementary particle physics with astrophysics. It is now possible to apply the laws governing small particles in the description of the nature of large particles hence the possibility of combining quantum mechanics with general relativity has been given out in detail. The equation for the first law of thermodynamics has also been generalized to $U = M_p c^2 - M_p (R_s^2 [E_o kT / h^2]) = M_p c^2 - M_p (R_s^2 E_o [4 \pi^2 / \hbar \tau])$ where τ defines the life time and $M_p = 2.1765 \times 10^{-8}$ Kg is the Planck mass. These results therefore show a clear future for the formulation of the unified law of all of physics.

23. Balungi's 2010 Research in Quantum Gravity

1.2. Gravity and Quantum mechanics

1.2.1. Gravity

By definition; to a person standing on the earth's surface, gravity is a force that pulls us towards the center of the earth. The force of gravity is realized when the phenomena being studied has a large mass. In other words the force of gravity is felt between two large bodies, that is, between the earth and the sun. Also the gravitational force is what keeps the earth in orbit around the sun and its strength increases as the distance between the two bodies' increases. For purposes of this book we disregard Einstein's General relativity theory/its mathematics and interpretations of gravity has gravitons. For purposes of this book we strongly reject any verbal and non verbal ideas about gravitons as these have lead to students who love the subject to be confused.

1.2.2. Quantum mechanics or Quantum theory

This is a mathematical description of subatomic particles and how they interact. It describes the strength of the electromagnetic force between two electrons by an exchange of quanta known as a photon. The quantum theory also explains the wave particle duality of matter and its interactions. For purposes of this book the langragian, laplace equations will not be used. The descriptions in this book are based on common sense and are not complicated in anyway.

To fully understand the theory of quantum gravity one needs to know the meaning of light. **What is light?** In a broad description, light is an electromagnetic wave made up of particles (photons) of discrete energy bundles like bullets shot from a gun. Light can also be bent when reflected towards a strong gravitational field. This definition is what is called a wave- particle duality.\

The light we have described above is the same light that is emitted from an atom whenever an electron in orbit around the nucleus jumps from one energy level to another. It is also the same light that is emitted by the sun. Whatever happens in the atom is the beginning to our development of the theory of quantum gravity, because like the earth's orbit around the sun it is the same mechanism by which the electron orbits the nucleus of an atom. The difference between the electron-nucleus system and the earth sun solar system is that in the atom the force that keeps an electron in orbit is the electromagnetic force while that of the earth's orbit is the gravitational pull. It has been said that, "the effects of gravity in an atom are so small and therefore negligible" but in this book they are introduced into an atom using a postulate that states; *two energies are possessed by an electron whenever it jumps from one energy level to another and that is, the energy of a photon and the potential gravitational energy possessed by an electron when moved from one state to another.* The two energies will never be equal since the speed of light of the emitted photon/electromagnetic wave will never be equal to the speed of an electron in orbit around the nucleus.

$$\frac{energy\ of\ a\ photon}{gravitational\ potential\ energy} = \frac{speed\ of\ light/photon}{speed\ of\ an\ electron}$$

By definition: the speed of light is related to the electric and magnetic fields by $c = \frac{E}{B}$, also the electric field is defined as the electric force per unit charge for an electron placed in vicinity of an electric field $E = \frac{F_e}{e}$, but the magnetic field in this case for a moving electron at a speed v is $B = \frac{F_b}{ve}$. The gravitational potential energy is the product of the gravitational force and the radius of orbit of an electron r $F_g\, r$. When the above definitions are taken into account then the energy of the emitted photon is related to the gravitational and electromagnetic force by

$$w_p = \frac{F_g F_e}{F_b} r \dots\dots\dots\dots\dots (1)$$

We have therefore hypothesized that the gravitational, the electric and magnetic force acting at a radius r of orbit of an electron around the nucleus of an atom are all relevant in the emission of a photon whenever an electron jumps from one energy level to another level.

1.2.3. The intensity of an electron in the electric field

The speed of a photon /light emitted from an atom due to the electric force is given by $c = \frac{F_e r^2}{n\hbar}$ and the angular momentum of an electron in orbit due to the gravitational force is $mvr = \frac{F_g r^2}{c}$. But since the speed of light doesnot change in all situations or by substituting the speed of light form equation $c = \frac{F_e r^2}{n\hbar}$ into the equation of the angular momentum, then the angular momentum of an electron is given by

$$mvr = \frac{F_g}{F_e} n\hbar \quad \dots\dots\dots(2)$$

If we treat the electron to be moving at a speed of light in the magnetic field then the power of an electron will be given as a product the magnetic force on an electron and the speed of light as $p = F_b c$, we can generalize the power of an electron on a quantum scale by making F_g the subject from equation (2) above and substituting it in equation (1). Take note that in this case the energy of a photon will be given by $W_p = h\frac{c}{\lambda}$. The power of an electron in orbit around the nucleus of an atom wll be given by,

$$p = \frac{2\pi r^2 \lambda m v F_e^2}{nh^2}$$

since an electron in orbit around the nucleus exbhits deBrogile wave properties and also moves in a spherical shape then the two assumptions hold that, if the wavelength $\lambda = \frac{h}{mv}$ and the surface area of a sphere is $A = 4\pi r^2$ and the intensity is power per unit area $I = \frac{P}{A}$, then the intensity of an electron on a quantum scale will be given by,

$$I_n = \frac{E^2 e^2}{2nh} \dots\dots\dots\dots\dots\dots\dots\dots\dots\dots\dots\dots\dots\dots\dots(3)$$

This implies that for each energy level the intensity of an electron is proportional to the square of its electric field.

However on a quantum gravity scale we can generate the intensity of a massive body in the gravitational field by taking an hypothesis that, if we treat the electromagnetic force Ee as an inverse square law we obtain the force on the particles in orbit as with the case for an electron in orbit around a nucleus at a distance r, then the electromagnetic force between two particles will be given by

$$Ee = \frac{n^2 \hbar c}{8\pi r^2}$$

For the schwarzichild's radius $r = \frac{Gm}{c^2}$, the electromagnetic force is given as $Ee = \frac{n^2 \hbar c^5}{8\pi G^2 m^2}$, substituting this expression into equation3 **then the intensity on a quantum gravitational scale** will be given by

$$I = \frac{n^3 \hbar c^{10}}{256\pi^3 G^4 m^4}$$

The intensity of an electromagnetic wave can also be deduced from eqn3 by assuming that at a level where the principle quantum number n is equal in magnitude to the fine structure constant or the electromagnetic coupling constant $\alpha = \frac{e^2}{4\pi \varepsilon_o \hbar c}$ then the intensity of an electromagnetic wave in the electric field and the combined electric and magnetic field is given by

$$I = 2\varepsilon_o E^2 c = \frac{EB}{\mu_o} \dots\dots\dots\dots\dots\dots\dots\dots\dots\dots\dots(4)$$

This implies that, the properties of an electromagnetic wave will only be present whenever the principle quantum number is the electromagnetic coupling constant or fine structure constant of value $n = \alpha = \frac{1}{137}$.

2.1. Black hole radiations/ Hawking radiations

By definition as of Wikipedia, a black hole is a mathematically defined region of space time exhibiting such a strong gravitational pull that no particle or electromagnetic radiation can escape from it. Many theories have been created to explain the properties of the black hole but the theory created here is far more different from the other theories although it may give the same results. Using a quite different approach towards solving a problem is efficient since it comes with it new predictions in the process which could have been hidden in other approaches. Below we try to present adhoc proofs-laws that may be of help in building our theory about black holes, note; these proofs can be derived mathematically from equations 1 up to 4 above but for purposes of simplicity they have been listed here below, however their derivations will be given in the last chapters of this book.

The laws or equations:

1) It is well known that the electric field is force per unit charge but here a generalized equation for an electric field created by an electron exhibiting wave properties in the nucleus of an atom in the gravitational field on a quantum scale is given by

$$E = \frac{1}{r}\sqrt{\frac{Gm^3 f}{2\hbar\varepsilon_o}} \quad \ldots(5)$$

Then the electric force in this case will be formulated as

$$F_1 = \frac{e}{r}\sqrt{\frac{Gm^3 f}{2\hbar\varepsilon_o}} \quad \ldots(6)$$

2) The surface area at a radius r of orbit of an electron of mass m around the nucleus of an atom in a wave like manner is given by

$$surface\ area(A) = \frac{\lambda\mu_o e^2}{m} \quad \ldots(7)$$

3) The time taken by the magnetic field B of an electron to pass through a given surface is

$$time(t) = \frac{\lambda \varepsilon_o AB}{e} \quad \text{...............(8)}$$

Note: the above expression is the same as Faraday's induction law.

4) The gravitational force acting on all matter in the universe or the modified gravitational force is given as

$$F_2 = \left(\frac{Gm^3}{r^2}\right)\left(\frac{e}{2B\lambda\hbar\varepsilon_o}\right) \quad \text{...............(9)}$$

The above formulas are important in deriving the formula for the temperature, entropy and the time taken by a black hole to evaporate as shown below;

2.1.1 Temperature of a black hole

It is known that the kinetic energy KE of molecules in the Boltzmann hypothesis is related to the temperature of the body in question in this case a black hole (in relation to the black body) by $KE = \varphi T$ where φ is Boltzmann's constant. The formula for the kinetic energy can be derived by using a hypothesis that the electromagnetic force – coulombs force is equal to eqn6 as

$$\frac{ke^2}{r^2} = \frac{e}{r}\sqrt{\frac{Gm^3 f}{2\hbar\varepsilon_o}}$$

On squaring both sides of the equation, cancelling like terms and taking into account that the frequency of an electron is $f = \frac{v}{\lambda}$, then the kinetic energy of an electron inside the black hole is given by

$$KE = \frac{\lambda\mu_o e^2}{A}\frac{c^3\hbar}{8\pi G m^2}$$

Since the surface area is given as from eqaution7 then **the kinetic energy of molecules** or particles (for an ideal gas) within the black hole will be given by

$$KE = \frac{c^3\hbar}{8\pi Gm} = T\varphi \quad \dots\dots\dots\dots\dots\dots\dots\dots(10)$$

Then from Boltzmann's relationship the temperature of the black hole is formulated as

$$T = \frac{c^3 \hbar}{8\pi G m \varphi} \ldots\ldots\ldots\ldots\ldots\ldots\ldots\ldots\ldots\ldots\ldots(11)$$

2.1.2 The entropy of the black hole

By definition entropy is a measure of disorder. To solve the entropy of black holes we shall consider a very complex argument about the entropy in question. We assume that the modified gravitational force given by equation 9 is identical to the modified electric field given by equation 6 as, $\left(\frac{Gm^3}{r^2}\right)\left(\frac{e}{2B\lambda\hbar\varepsilon_o}\right) \equiv \frac{e}{r}\sqrt{\frac{Gm^3 f}{2\hbar\varepsilon_o}}$ in otherwise the two forces are equal but opposite. Then squaring both sides of the equation and multiplying through by Gc^5 one obtains a new relation of forces on both sides given as

$$\frac{tc^7}{16\pi G^2 m} = \frac{Ac^6}{32\pi rmG^2}$$

Both the left and right hand side represent a force. From the left hand side t is the expression of time given by $t = \frac{\hbar e^2}{2m^3 c^2 G\varepsilon_o}$. Note: the left hand side force is the pull of matter inside the black hole while the right hand side force is the force acting on particles or matter at the surface of the black hole.

Since the heat is the product of the force on a particle and the distance r from the centre of the black hole, then using the force on the right hand side of the above equation the heat will be given by

$$Q = \frac{Ac^6}{32\pi m G^2}$$

Remember the temperature of the black hole is also known from equation9 and by definition the entropy of the system is the change in heat per unit temperature $\frac{Q}{T}$, then the entropy of the black hole will be given by

$$S = \frac{A\varphi c^3}{4G\hbar} \quad\quad\quad\quad\quad (12)$$

This implies that the entropy of a black hole is proportional to its surface area.

2.1.3 The time taken by a black hole to evaporate

Assuming that particles that formed a black hole are moving away or are separating from it after a given time of its existence, if we measure the relative speed of these particles in relation to the energy they carry we obtain a relation ship given by

$$\frac{v^2}{c^2} = \frac{8\pi G}{c^2} \left(\frac{W}{8\pi r}\right) \quad \dots\dots\dots\dots\dots\dots\dots\dots\dots\dots\dots\dots\dots\dots(13)$$

Where v is the velocity of these particles as measured relative to the speed of light c and W is the energy carried by the particles as they move away from the centre of the black hole at a distance r.

If we let the force causing the particles to separate from the black hole be given as $\frac{Gm^3 e}{2r\lambda B \hbar \varepsilon_o} \frac{v}{c}$, then the energy of these particles will be given by

$$W = \frac{Gm^3 e}{2r\lambda B \hbar \varepsilon_o} \frac{v}{c}$$

Substituting this in equation 11, we obtain a relation ship of time as given by the law 3 of equation 8 as

$$t = \frac{v^2}{c^2}\left(\frac{\pi G^2 m^3}{\hbar c^4}\right)$$

The velocity of the particles in the astronomical lab will be measured as v= 4.193E6 m/s and since the speed of light is a constant then the time taken by a black hole to evaporate is given by

$$t = \frac{5120\,\pi G^2 m^3}{\hbar c^4} \quad \dotfill (14)$$

Bibliography

Balungi Francis, (2010) "A hypothetical investigation into the realm of the microscopic and macroscopic universes beyond the standard model" general physics arXiv:1002.2287v1 [physics.gen-ph]

Milgrom, M. (1983). *"A modification of the Newtonian dynamics as a possible alternative to the hidden mass hypothesis"* Astrophysical journal. 270: 365-370

Sakharov, A.D (1967). *"Vacuum Quantum Fluctuations in curved space and the theory of Gravitation"* General Relativity and Gravitation Vol. 32, No.2, 2000 (Translated)

Einstein, Albert (1916). "The Foundation of the General Theory of Relativity". Annalen der Physik. **354** (7): 769. *Bibcode:1916AnP...354..769E.doi:10.1002/andp.19163540702.* Archived from *the original (PDF)* on 2012-02-06.

Isaac Newton, The Third Book of *Opticks* (2nd ed. 1718).

Albert Einstein's 'First' Paper (1894 or 1895), http://www.straco.ch/papers/Einstein%20First%20Paper.pdf

Einstein, Albert: "Ether and the Theory of Relativity" (1920), republished in *Sidelights on Relativity* (Methuen, London, 1922)

Laughlin, Robert B. (2005). *A Different Universe: Reinventing Physics from the Bottom Down*. NY, NY: Basic Books. pp. *120–121. ISBN 978-0-465-03828-2.*

Wetterich, C. *"Quintessence --a fifth force from variation of the fundamental scale" (PDF)*. Heidelberg University.

Dvali, Gia; Zaldarriaga, Matias (2002). *"Changing α With Time: Implications For Fifth-Force-Type Experiments And Quintessence" (PDF)*. Physical Review Letters. **88** (9): 091303. *arXiv:hep-ph/0108217. Bibcode:2002PhRvL..88i1303D. doi:10.1103/PhysRevLett.88.091303. PMID 11863992.*

Cicoli, Michele; Pedro, Francisco G.; Tasinato, Gianmassimo (23 July 2012). "Natural Quintessence in String Theory" – via arXiv.org.

Hawking, Stephen (1975). "Particle Creation by Black Holes". Commun. Math. Phys. 43 (3): 199–220. Bibcode:1975CMaPh..43..199H.

Hawking, S. W. (1974). "Black hole explosions?". Nature.248(5443):30–31.

Bibcode:1974Natur.248...30H.doi:10.1038/248030a0.

Carlo Rovelli (2003) "Quantum Gravity" Draft of the Book Pdf

Some few texts used are from Wikipedia https://creativecommons.org/licenses/by-sa/3.0/

D. N. Page, Phys. Rev. D 13, 198 (1976).

C. Gao and Y.Lu, Pulsations of a black hole in LQG (2012) arXiv:1706.08009v3

A.H. Chamseddine and V.Mukhanov, Non singular black hole (2016) arXiv 1612.05861v1

M.Bojowald and G.M.Paily, A no-singularity scenario in LQG (2012) arXiv: 1206.5765v1

P.Singh, class.Quant.Grav,26,125005(2009), arXiv:0901.2750

P.Singh and F.Vidotto, Phys.Rev, D83,064027(2011) arXiv:1012.1307

C.Rovelli and F.Vidotto, Phy. Rev,111(9) 091303(2013) arXiv:1307.3228v2

M.Bojowald, Initial conditions for a universe, Gravity Research Foundation (2003)

A.Ashtekar, Singularity Resolution in Loop Quantum Cosmology (2008) arXiv:0812.4703v1

J.Brunneumann and T.Thiemann, On singularity avoidance in Loop Quantum Gravity (2005) arXiv:0505032v1

L.Modesto, Disappearence of the Black hole singularity in Quantum gravity (2004) arXiv:0407097v2

Mikhailov, A.A. (1959).Mon. Not. Roy. Astron. Soc.,119, 593.

P. Merat etal.(1974). Astron & Astrophys 32, 471-475

Trempler, R.J. (1956).Helv. Phys. Acta, Suppl.,IV, 106.

Trempler, R.J. (1932). " The deflection of light in the sun's gravitational field "Astronomical society of the pacific 167

Einstein, A. (1916).Ann. d. Phys.,49, 769; (1923).The Principle of Relativity, (translators Perret, W. and Jeffery, G.B.), (Dover Publications, Inc., New York), pp. 109–164.

Von Klüber, H. (1960). InVistas in Astronomy, Vol. 3, pp. 47–77.

K. Hentschel (1992). Erwin Finlay Freundlich and testing Einstein theory of relativity, Communicated by J.D. North

Muhleman, D.O., Ekers, R.D. and Fomalont, E.B. (1970).Phys. Rev. Lett.,24, 1377

Mikhailov, A.A. (1956).Astron. Zh.,33, 912.

Dyson, F.W., Eddington, A.S. and Davidson, C. (1920).Phil. Trans. Roy. sog., A220, 291

Chant, C.A. and Young, R.K. (1924).Publ. Dom. Astron. Obs.,2, 275.

Campbell, W.W. and Trumbler, R.J. (1923).Lick Obs. Bull.,11, 41.

Freundlich, E.F., von Klüber, H. and von Brunn, A. (1931).Abhandl. Preuss. Akad. Wiss. Berlin, Phys. Math. Kl., No.l;Z. Astrophys.,3, 171

Mikhailov, A.A. (1949).Expeditions to Observe the Total Solar Eclipse of 21 September, 1941, (report), (ed. Fesenkov, V.G.), (Publications of the Academy of Sciences, U.S.S.R.), pp. 337–351.

S.P. Martin, in Perspectives on Supersymmetry , edited by G.L. Kane (World Scientific, Singapore, 1998) pp. 1–98; and a longer archive version in hep-ph/9709356; I.J.R. Aitchison, hep-ph/0505105.

Mara Beller, Quantum Dialogue: The Making of a Revolution. University of Chicago Press, Chicago, 2001.

Morrison, Philp: "The Neutrino, scientific American, Vol 194,no.1 (1956),pp.58-68.

R. Haag, J. T. Lopuszanski and M. Sohnius, Nucl. Phys. B88, 257 (1975) S.R. Coleman and J. Mandula, Phys.Rev. 159 (1967) 1251.

H.P. Nilles, Phys. Reports 110, 1 (1984).

P. Nath, R. Arnowitt, and A.H. Chamseddine, Applied N = 1 Supergravity (World Scientific, Singapore, 1984).

S.P. Martin, in Perspectives on Supersymmetry, edited by G.L. Kane (World Scientific, Singapore, 1998) pp. 1–98; and a longer archive version in hep-ph/9709356; I.J.R. Aitchison, hep-ph/0505105.

S. Weinberg, The Quantum Theory of Fields, VolumeIII: Supersymmetry (Cambridge University Press, Cambridge,UK, 2000).

E. Witten, Nucl. Phys. B188, 513 (1981).

S. Dimopoulos and H. Georgi, Nucl. Phys. B193, 150(1981).

N. Sakai, Z. Phys. C11, 153 (1981);R.K. Kaul, Phys. Lett. 109B, 19 (1982).

L. Susskind, Phys. Reports 104, 181 (1984).

L. Girardello and M. Grisaru, Nucl. Phys. B194, 65(1982); L.J. Hall and L. Randall,

Phys. Rev. Lett. 65, 2939(1990);I. Jack and D.R.T. Jones, Phys. Lett. B457, 101 (1999).

For a review, see N. Polonsky, Supersymmetry: Structureand phenomena. Extensions of the standard model, Lect.Notes Phys. M68, 1 (2001).

G. Bertone, D. Hooper and J. Silk, Phys. Reports 405, 279 (2005).

G. Jungman, M. Kamionkowski, and K. Griest, Phys. Reports 267, 195 (1996).

V. Agrawal, S.M. Barr, J.F. Donoghue and D. Seckel, Phys. Rev. D57, 5480 (1998).

N. Arkani-Hamed and S. Dimopoulos, JHEP 0506, 073(2005); G.F. Giudice and A. Romanino, Nucl. Phys. B699, 65(2004) [erratum: B706, 65 (2005)]. July 27, 2006 11:28

en.wikipedia.org/wiki/Supersymmetry - 52k - Cached - Similar pages

en.wikipedia.org/wiki/Grand_unification_theory - 39k - Cached - Similar pages

In cosmology, the Planck epoch (or Planck era), named after Max Planck, is the earliest period of time in the history of the universe, en.wikipedia.org/wiki/**Planck_epoch** - 23k - Cached - Similar pages

L. Shapiro and J. Sol`a, Phys. Lett. B 530, 10 (2002);

E. V.Gorbar and I. L. Shapiro, JHEP 02, 021 (2003); A. M. Pelinson,

L. Shapiro, and F. I. Takakura, Nucl. Phys. B 648, 417 (2003).

A. Starobinsky, Phys. Lett. B 91, 99 (1980).

G. F. R. Ellis, J. Murugan, and C. G. Tsagas, Class. Quant. Grav.21, 233 (2004).

H. V. Peiris et al., Astrophys. J. Suppl. 148, 213 (2003).

D. N. Spergel et al., astro-ph/0603449.

Vilenkin, Phys. Rev. D 32, 2511 (1985).

A. Starobinsky, Pis'ma Astron. Zh 9, 579 (1983).

A.H. Guth, Phys. Rev. D23, 347 (1981).

A.D. Linde, Phys. Lett. B108, 389 (1982); A. Albrecht, P.J. Steinhardt, Phys.Rev. Lett. 48, 1220 (1982).

A.D. Linde, Phys Lett. B129, 177 (1983).

N. Makino, M. Sasaki, Prog. Theor. Phys. 86, 103 (1991);

D. Kaiser, Phys. Rev.D52, 4295 (1995).

H. Goldberg, Phys. Rev. Lett. 50, 1419 (1983).

E. Kolb and M. Turner, The Early Universe (Addison-Wesley, Reading, MA,1990).

W. Garretson and E. Carlson, Phys. Lett. B 315, 232(1993); H. Goldberg, hep-ph/0003197.

Eddington, A. S., The Internal Constitution of the Stars (Cambridge University Press, England,1926), p. 16

Duncan R .C. & Thompson C., Ap.J.392, L 9 (1992).

Thompson , C, Duncan , R .C ., Woods , P., Kouveliotou , C ., Finger , M.H. & van Parad ij s , J .,ApJ, submitted , astro-ph /9908086, (2000).

Schwinger , J .,Phys. Rev.73, 416L (1948)

Carlip, S.: Quantum gravity: a progress report. Rept. Prog. Phys. 64, 885 (2001).arXiv:gr-qc/0108040

Kerr,R.P.: Gravitational field of a spinning mass as an example of algebraically special metrics.

Phys. Rev. Lett. 11, 237–238 (1963)

Bekenstein, J.: Black holes and the second law. Lett. Nuovo Cim. 4, 737–740 (1972)

Bardeen, J.M., Carter, B., Hawking, S.: The four laws of black hole mechanics. Commun.

Math. Phys. 31, 161–170 (1973)

Tolman, R.: Relativity, Thermodynamics, and Cosmology. Dover Books on Physics Series.

Dover Publications, New York (1987)

Oppenheimer, J., Volkoff, G.: On massive neutron cores. Phys. Rev. 55, 374–381 (1939)

Tolman, R.C.: Static solutions of einstein's field equations for spheres of fluid, Phys. Rev. 55, 364–373 (1939)

Zel'dovich Y.B.: Zh. Eksp. Teoret. Fiz.41, 1609 (1961)

Bondi, H.: Massive spheres in general relativity. Proc. Roy. Soc. Lond. A281, 303–317 (1964)

Sorkin, R.D., Wald, R.M., Zhang, Z.J.: Entropy of selfgravitating radiation. Gen. Rel. Grav. 1127–1146 (1981)

Newman, E.T., Couch, R., Chinnapared, K., Exton, A., Prakash, A., et al.: Metric of a rotating,charged mass. J. Math. Phys. 6, 918–919 (1965)

Ginzburg, V., Ozernoi, L.: Sov. Phys. JETP 20, 689 (1965)

Doroshkevich, A., Zel'dovich, Y., Novikov I.: Gravitational collapse of nonsymmetric and rotating masses, JETP 49 (1965)

Israel, W.: Event horizons in static vacuum space-times. Phys. Rev. 164, 1776–1779 (1967)

Israel,W.: Event horizons in static electrovac space-times. Commun. Math. Phys. 8, 245–260 (1968)

Loop quantum gravity does provide such a prediction [363, 364], and it disagrees with the semiclassical

Carter, B.: Axisymmetric black hole has only two degrees of freedom. Phys. Rev. Lett. 26, 331–333(1971)

Penrose, R.: Gravitational collapse: the role of general relativity. Riv. Nuovo Cim. 1, 252–276 (1969)

Christodoulou, D.: Reversible and irreversible transformations in black hole physics. Phys. Rev. Lett. 25, 1596–1597 (1970)

Christodoulou, D., Ruffini, R.: Reversible transformations of a charged black hole. Phys. Rev. D4, 3552–3555 (1971)

Hawking, S.: Particle creation by black holes. Commun. Math. Phys. 43, 199–220 (1975)

Klein, O.: Die reflexion von elektronen an einem potential sprung nach der relativistischen dynamik von dirac. Z. Phys. 53, 157 (1929)

Wald, R.M.: General Relativity. The University of Chicago Press, Chicago (1984)

Hawking, S.W.: Black hole explosions. Nature 248, 30–31 (1974)

Hawking, S., Ellis, G.: The large scale structure of space-time. Cambridge University Press, Cambridge (1973)

Carter, B.: Black hole equilibrium states, In Black Holes— Les astres occlus. Gordon and Breach Science Publishers, (1973)

Hawking, S.W.: Gravitational radiation from colliding black holes. Phys. Rev. Lett. 26, 1344– 1346 (1971)

Hawking, S.: Black holes in general relativity. Commun. Math. Phys. 25, 152–166 (1972)

Bekenstein, J.: Extraction of energy and charge from a black hole. Phys. Rev. D7, 949–953 (1973)

Bekenstein, J.D.: Black holes and entropy. Phys. Rev. D7, 2333–2346 (1973)

Hawking, S.: Quantum gravity and path integrals. Phys. Rev. D18, 1747–1753 (1978)

Gross, D.J., Perry, M.J., Yaffe, L.G.: Instability of flat space at finite temperature. Phys. Rev. D25, 330–355 (1982)

Unruh, W.G., Wald, R.M.: What happens when an accelerating observer detects a rindler particle. Phys. Rev. D29, 1047–1056 (1984)

Bekenstein, J.D.: Auniversal upper bound on the entropy to energy ratio for bounded systems. Phys. Rev. D23, 287 (1981)

Unruh,W.,Wald, R.M.: Acceleration radiation and generalized second law of thermodynamics. Phys. Rev. D25, 942–958 (1982)

Unruh, W., Wald, R.M.: Entropy bounds, acceleration radiation, and the generalized second law. Phys. Rev. D27, 2271–2276 (1983)

Image : MPI for gravitational physics / W.Benger-zib

Tomilin,K.A., (1999). "Natural Systems Of Units: To The Centenary Aniniversary Of   The Planck Systems", 287-296

Sivaram, C. (2007). "What Is Special About the Planck Mass"? arXiv:0707.0058v1

H. Georgi and S.L. Glahow. (1974) "Unity Of All Elementary-Particle Forces". Phys. Rev. Letters 32, 438

Luigi Maxmilian Caligiuri, Amrit Sorli. Gravity Originates from Variable Energy Density of Quantum Vacuum. American Journal of Modern Physics. Vol. 3, No. 3, 2014, pp. 118-128. doi: 10.11648/j.ajmp.20140303.11

Philip J. Tattersall,(2018) Quantum Vacuum Energy and the Emergence of Gravity. doi:10.5539/apr.v10n2p1

H. E. Puthoff (1989) Gravity as a zero-point-fluctuation force PHYSICAL REVIEW A VOLUME 39, NUMBER 5

Balungi Francis (2018) "Quantum Gravity in a Nutshell1" Book.

E.Verlinde (2016) Emergent Gravity and the Dark Universe, arXiv:1611.02269v2[hep-th]

S.Hossenfelder (2018) The Redshift-Dependence of Radial Acceleration: Modified gravity versus particle dark matter, arXiv:1803.08683v1[gr-qc]

Robert J. Scherrer (2004) Purely kinetic k-essence as unified dark matter, arXiv:astro-ph/0402316v3

J.S.Farnes (2018), Aunifying theory of dark energy and dark matter: Negative masses and matter creation within a modified ΛCDM framework, arXiv:1712.07962v2[physics.gen-ph]

Gustav M Obermair (2013), Primordial Planck mass black holes (PPMBHs) as candidates for dark matter? J. Phys:conf.Ser.442012066

V.Cooray etal...(2017), An alternative approach to estimate the vacuum energy density of free space, doi:10.20944/preprints201707.0048.v1

M.Milgrom, (1983) A modification of the Newtonian dynamics: Implications for galaxies, Astrophys.J.270, 371.

Acknowledgments

This book would never have been completed without the patience and dedication of my wife, Wanyana Ritah. She performed the wonderful and difficult task of editing major parts of the book and helped in researching many details necessary to complete it.

I wish to thank several colleagues for their help and extensive comments on the manuscript, including Lee Smolin, Carlo Rovelli, Sabine Hossenfelder, Jim Baggot and Viktor Toth. I also thank my colleagues Harvey Brown, Paul Frampton, Stacy McGaugh, and Lee Smolin for helpful comments. I particularly thank a total of 200 online physics friends and SUSP science foundation members, for a careful reading of the manuscript. Many graduate students have contributed over the years to developing my pressure vacuum theory of gravity.

I also wish to thank my editors, at Think Physics for their enthusiasm and support. Finally, I thank our family for their patience, love, and support during the four years of working on this book.

About the Author

Balungi Francis was born in Kampala, Uganda, to a single poor mother, grew up in Kawempe, and later joined Makerere Universty in 2006, graduating with a Bachelor Science degree in Land Surveying in 2010. For four years he taught in Kampala City high schools, majoring in the fields of Gravitation and Quantum Physics. His first book, "Mathematical Foundation of the Quantum theory of Gravity," won the Young Kampala Innovative Prize and was mentioned in the African Next Einstein Book Prize (ANE).

He has spent over 15years researching and discovering connections in physics, mathematics, geometry, cosmology, quantum mechanics, gravity, in addition to astrophysics, unified physics and geographical information systems . These studies led to his groundbreaking theories, published papers, books and patented inventions in the science of Quantum Gravity, which have received worldwide recognition.

From these discoveries, Balungi founded the SUSP (Solutions to the Unsolved Scientific Problems) Project Foundation in 2004 - now known as the SUSP Science Foundation. As its current Director of Research, Balungi leads physicists, mathematicians and engineers in exploring Quantum Gravity principles and their implications in our world today and for future generations.b

Balungi launched the Visionary School of Quantum Gravity in 2016 in order to bring the learning and community further together. It's the first and only Quantum Gravity physics program of its kind, educating thousands of students from over 80 countries.

The book "Quantum Gravity in a Nutshell1", a most recommend book in quantum gravity research, was produced based on Balungi's discoveries and their potential for generations to come. Balungi is currently guiding the Foundation, speaking to audiences worldwide, and continuing his groundbreaking research.

Glossary

Absolute space and time—the Newtonian concepts of space and time, in which space is independent of the material bodies within it, and time flows at the same rate throughout the universe without regard to the locations of different observers and their experience of "now."

Acceleration—the rate at which the speed or velocity of a body changes.

Accelerating universe—the discovery in 1998, through data from very distant supernovae, that the expansion of the universe in the wake of the big bang is not slowing down, but is actually speeding up at this point in its history; groups of astronomers in California and Australia independently discovered that the light from the supernovae appears dimmer than would be expected if the universe were slowing down.

Action—the mathematical expression used to describe a physical system by requiring only the knowledge of the initial and final states of the system; the values of the physical variables at all intermediate states are determined by minimizing the action.

Anthropic principle—the idea that our existence in the universe imposes constraints on its properties; an extreme version claims that we owe our existence to this principle.

Asymptotic freedom (or safety)—a property of quantum field theory in which the strength of the coupling between elementary particles vanishes with increasing energy and/or decreasing distance, such that the elementary particles approach free particles with no external forces acting on them; moreover for decreasing energy and/or increasing distance between the particles, the strength of the particle force increases indefinitely.

Baryon—a subatomic particle composed of three quarks, such as the proton and neutron.

Big bang theory—the theory that the universe began with a violent explosion of spacetime, and that matter and energy originated from an infinitely small and dense point.

Big crunch—similar to the big bang, this idea postulates an end to the universe in a singularity.

Binary stars—a common astrophysical system in which two stars rotate around each other; also called a "double star."

Blackbody—a physical system that absorbs all radiation that hits it, and emits characteristic radiation energy depending upon temperature; the concept of blackbodies is useful, among other things, in learning the temperature of stars.

Black hole—created when a dying star collapses to a singular point, concealed by an "event horizon;" the black hole is so dense and has such strong gravity that nothing, including light, can escape it; black holes are predicted by general relativity, and though they cannot be "seen," several have been inferred from astronomical observations of binary stars and massive collapsed stars at the centers of galaxies.

Boson—a particle with integer spin, such as photons, mesons, and gravitons, which carries the forces between fermions.

Brane—shortened from "membrane," a higher-dimensional extension of a onedimensional string.

Cassini spacecraft—NASA mission to Saturn, launched in 1997, that in addition to making detailed studies of Saturn and its moons, determined a bound on the variations of Newton's gravitational constant with time.

Causality—the concept that every event has in its past events that caused it, but no event can play a role in causing events in its past.

Classical theory—a physical theory, such as Newton's gravity theory or Einstein's general relativity, that is concerned with the macroscopic universe, as opposed to theories concerning events at the submicroscopic level such as quantum mechanics and the standard model of particle physics.

Copernican revolution—the paradigm shift begun by Nicolaus Copernicus in the early sixteenth century, when he identified the sun, rather than the Earth, as the center of the known universe.

Cosmic microwave background (CMB)—the first significant evidence for the big bang theory; initially found in 1964 and studied further by NASA teams in 1989 and the early 2000s, the CMB is a smooth signature of microwaves everywhere in the sky, representing the "afterglow"of the big bang: Infrared light produced about 400,000 years after the big bang had redshifted through the stretching of spacetime during fourteen billion years of expansion to the microwave part of the electromagnetic spectrum, revealing a great deal of information about the early universe.

Cosmological constant—a mathematical term that Einstein inserted into his gravity field equations in 1917 to keep the universe static and eternal; although he later regretted this and called it his "biggest blunder," cosmologists today still use the cosmological constant, and some equate it with the mysterious dark energy.

Coupling constant—a term that indicates the strength of an interaction between particles or fields; electric charge and Newton's gravitational constant are coupling constants.

Crystalline spheres—concentric transparent spheres in ancient Greek cosmology that held the moon, sun, planets, and stars in place and made them revolve around the Earth; they were part of the western conception of the universe until the Renaissance.

Curvature—the deviation from a Euclidean spacetime due to the warping of the geometry by massive bodies.

Dark energy—a mysterious form of energy that has been associated with negative pressure vacuum energy and Einstein's cosmological constant; it is hypothesized to explain the data on the accelerating expansion of the universe; according to the standard model, the dark energy, which is spread uniformlythroughout the universe, makes up about 70 percent of the total mass and energy content of the universe.

Dark matter—invisible, not-yet-detected, unknown particles of matter, representing about 30 percent of the total mass of matter according to the standard model; its presence is necessary if Newton's and Einstein's gravity theories are to fit data from galaxies, clusters of galaxies, and cosmology; together, darkmatter and dark energy mean that 96 percent of the matter and energy in the universe is invisible.

Deferent—in the ancient Ptolemaic concept of the universe, a large circle representing the orbit of a planet around the Earth.

Doppler principle or **Doppler effect**—the discovery by the nineteenth-century Austrian scientist Christian Doppler that when sound or light waves are moving toward an observer, the apparent frequency of the waves will be shortened, while if they are moving away from an observer, they will be lengthened; in astronomy this means that the light emitted by galaxies moving away from us is redshifted, and that from nearby galaxies moving toward us is blueshifted.

Dwarf galaxy—a small galaxy (containing several billion stars) orbiting a larger galaxy; the Milky Way has over a dozen dwarf galaxies as companions, including the Large Magellanic Cloud and Small Magellanic Cloud.

Dynamics—the physics of matter in motion.

Electromagnetism—the unified force of electricity and magnetism, discovered to be the same phenomenon by Michael Faraday and James Clerk Maxwell in the nineteenth century.

Electromagnetic radiation—a term for wave motion of electromagnetic fields which propagate with the speed of light—300,000 kilometers per second—and differ only in wavelength; this includes visible light, ultraviolet light, infrared radiation,

X-rays, gamma rays, and radio waves.

Electron—an elementary particle carrying negative charge that orbits the nucleus of an atom.

Eötvös experiments—torsion balance experiments performed by Hungarian Count Roland von Eötvös in the late nineteenth and early twentieth centuries that showed that inertial and gravitational mass were the same to one part in 1011; this was a more accurate determination of the equivalence principle than results achieved by Isaac Newton and, later, Friedrich Wilhelm Bessel.

Epicycle—in the Ptolemaic universe, a pattern of small circles traced out by a planet at the edge of its "deferent" as it orbited the Earth; this was how the Greeks accounted for the apparent retrograde motions of the planets.

Equivalence principle—the phenomenon first noted by Galileo that bodies falling in a gravitational field fall at the same rate, independent of their weight and composition; Einstein extended the principle to show that gravitation is identical (equivalent) to acceleration.

Escape velocity—the speed at which a body must travel in order to escape a strong gravitational field; rockets fired into orbits around the Earth have calculated escape velocities, as do galaxies at the periphery of galaxy clusters.

Ether (or aether)—a substance whose origins were in the Greek concept of "quintessence," the ether was the medium through which energy and matter moved, something more than a vacuum and less than air; in the late nineteenth century the Michelson-Morley experiment disproved the existence of the ether.

Euclidean geometry—plane geometry developed by the third-century bc Greek mathematician Euclid; in this geometry, parallel lines never meet.

Fermion—a particle with half-integer spin, like protons and electrons, that make up matter.

Field—a physical term describing the forces between massive bodies in gravity and electric charges in electromagnetism; Michael Faraday discovered the concept of field when studying magnetic conductors.

Field equations—differential equations describing the physical properties of interacting massive particles in gravity and electric charges in electromagnetism; Maxwell's equations for electromagnetism and Einstein's equations of gravity are prominent examples in physics.

Fifth force or **"skew" force**—a new force in MOG that has the effect of modifying gravity over limited length scales; it is carried by a particle with mass called the phion.

Fine-tuning—the unnatural cancellation of two or more large numbers involving an absurd number of decimal places, when one is attempting to explain a physical phenomenon; this signals that a true understanding of the physical phenomenon has not been achieved.

Fixed stars—an ancient Greek concept in which all the stars were static in the sky, and moved around the Earth on a distant crystalline sphere.

Frame of reference—the three spatial coordinates and one time coordinate that an observer uses to denote the position of a particle in space and time.

Galaxy—organized group of hundreds of billions of stars, such as our Milky Way.

Galaxy cluster—many galaxies held together by mutual gravity but not in as organized a fashion as stars within a single galaxy.

Galaxy rotation curve—a plot of the Doppler shift data recording the observed velocities of stars in galaxies; those stars at the periphery of giant spiral galaxies are observed to be going faster than they "should be" according to Newton's and Einstein's gravity theories.

General relativity—Einstein's revolutionary gravity theory, created in 1916 from a mathematical generalization of his theory of special relativity; it changed our concept of gravity from Newton's universal force to the warping of the geometry of spacetime in the presence of matter and energy.

Geodesic—the shortest path between two neighboring points, which is a straight line in Euclidian geometry, and a unique curved path in four-dimensional spacetime.

Globular cluster—a relatively small, dense system of up to millions of stars occurring commonly in galaxies.

Gravitational lensing—the bending of light by the curvature of spacetime; galaxies and clusters of galaxies act as lenses, distorting the images of distant bright galaxies or quasars as the light passes through or near them.

Gravitational mass—the active mass of a body that produces a gravitational force on other bodies.

Gravitational waves—ripples in the curvature of spacetime predicted by general relativity; although any accelerating body can produce gravitational radiation or waves, those that could be detected by experiments would be caused by cataclysmic cosmic events.

Graviton—the hypothetical smallest packet of gravitational energy, comparable to the photon for electromagnetic energy; the graviton has not yet been seen experimentally.

Group (in mathematics)—in abstract algebra, a set that obeys a binary operation that satisfies certain axioms; for example, the property of addition of integers makes a group; the branch of mathematics that studies groups is called group theory.

Hadron—a generic word for fermion particles that undergo strong nuclear interactions.

Hamiltonian—an alternative way of deriving the differential equations of motion for a physical system using the calculus of variations; Hamilton's principle is also called the "principle of stationary action" and was originally formulated by Sir William Rowan Hamilton for classical mechanics; the principle applies to classical fields such as the gravitational and electromagnetic fields, and has had important applications in quantum mechanics and quantum field theory.

Homogeneous—in cosmology, when the universe appears the same to all observers, no matter where they are in the universe.

Inertia—the tendency of a body to remain in uniform motion once it is moving, and to stay at rest if it is at rest; Galileo discovered the law of inertia in the early seventeenth century.

Inertial mass—the mass of a body that resists an external force; since Newton, it has been known experimentally that inertial and gravitational mass are equal; Einstein used this equivalence of inertial and gravitational mass to postulate his equivalence principle, which was a cornerstone of his gravity theory.

Inflation theory—a theory proposed by Alan Guth and others to resolve the flatness, horizon, and homogeneity problems in the standard big bang model; the very early universe is pictured as expanding exponentially fast in a fraction of a second.

Interferometry—the use of two or more telescopes, which in combination create a receiver in effect as large as the distance between them; radio astronomy in particular makes use of interferometry.

Inverse square law—discovered by Newton, based on earlier work by Kepler, this law states that the force of gravity between two massive bodies or point particles decreases as the inverse square of the distance between them.

Isotropic—in cosmology, when the universe looks the same to an observer, no matter in which direction she looks.

Kelvin temperature scale—designed by Lord Kelvin (William Thomson) in the mid-1800s to measure very cold temperatures, its starting point is absolute zero, the coldest possible temperature in the universe, corresponding to –

273.15 degrees Celsius; water's freezing point is 273.15K (0°C), while its boiling point is 373.15K (100°C).

Lagrange points—discovered by the Italian-French mathematician Joseph-Louis Lagrange, these five special points are in the vicinity of two orbiting masses where a third, smaller mass can orbit at a fixed distance from the larger masses; at the Lagrange points, the gravitational pull of the two large masses precisely equals the centripetal force required to keep the third body, such as a satellite, in a bound orbit; three of the Lagrange points are unstable, two are stable.

Lagrangian—named after Joseph-Louis Lagrange, and denoted by L, this mathematical expression summarizes the dynamical properties of a physical system; it is defined in classical mechanics as the kinetic energy T minus the potential energy V; the equations of motion of a system of particles may be derived from the Euler-Lagrange equations, a family of partial differential equations.

Light cone—a mathematical means of expressing past, present, and future space and time in terms of spacetime geometry; in four-dimensional Minkowski spacetime, the light rays emanating from or arriving at an event separate spacetime into a past cone and a future cone which meet at a point corresponding to the event.

Lorentz transformations—

mathematical transformations from one inertial frame of reference to another such that the laws of physics remain the same; named after Hendrik Lorentz, who developed them in 1904, these transformations form the basic mathematical equations underlying special relativity.

Mercury anomaly—a phenomenon in which the perihelion of Mercury's orbit advances more rapidly than predicted by Newton's equations of gravity; when Einstein showed that his gravity theory predicted the anomalous precession, it was the first empirical evidence that general relativity might be correct.

Meson—a short-lived boson composed of a quark and an antiquark, believed to bind protons and neutrons together in the atomic nucleus.

Metric tensor—mathematical symmetric tensor coefficients that determine the infinitesimal distance between two points in spacetime; in effect the metric tensor distinguishes between Euclidean and non-Euclidean geometry.

Michelson-Morley experiment—1887 experiment by Albert Michelson and Edward Morley that proved that the ether did not exist; beams of light traveling in the same direction, and in the perpendicular direction, as the supposed ether showed no difference in speed or arrival time at their destination.

Milky Way—the spiral galaxy that contains our solar system.

Minkowski spacetime—the geometrically flat spacetime, with no gravitational effects, first described by the Swiss mathematician Hermann Minkowski; it became the setting of Einstein's theory of gravity.

MOG—my relativistic modified theory of gravitation, which generalizes Einstein's general relativity; MOG stands for "Modified Gravity."

MOND—a modification of Newtonian gravity published by Mordehai Milgrom in 1983; this is a nonrelativistic phenomenological model used to describe rotational velocity curves of galaxies; MOND stands for "Modified Newtonian Dynamics."

Neutrino—an elementary particle with zero electric charge; very difficult to detect, it is created in radioactive decays and is able to pass through matter almost undisturbed; it is considered to have a tiny mass that has not yet been accurately measured.

Neutron—an elementary and electrically neutral particle found in the atomic nucleus, and having about the same mass as the proton.

Nuclear force—another name for the strong force that binds protons and neutrons together in the atomic nucleus.

Nucleon—a generic name for a proton or neutron within the atomic nucleus.

Neutron star—the collapsed core of a star that remains after a supernova explosion; it is extremely dense, relatively small, and composed of neutrons.

Newton's gravitational constant—the constant of proportionality, G, which occurs in the Newtonian law of gravitation, and says that the attractive force between two bodies is proportional to the product of their masses and inversely proportional to the square of the distance between them; its numerical value is: $G = 6.67428 \pm 0.00067 \times 10^{-11}$ m3 kg–1 s–2.

Nonsymmetric field theory (Einstein)—a mathematical description of the geometry of spacetime based on a metric tensor that has both a symmetric part and an antisymmetric part; Einstein used this geometry to formulate a unified field

theory of gravitation and electromagnetism.

Nonsymmetric Gravitation Theory (NGT)—my generalization of Einstein's purely gravitation theory (general relativity) that introduces the antisymmetric field as an extra component of the gravitational field; mathematically speaking, the nonsymmetric field structure is described by a non-Riemannian geometry.

Parallax—the apparent movement of a nearer object relative to a distant background when one views the object from two different positions; used with triangulation for measuring distances in astronomy.

Paradigm shift—a revolutionary change in belief, popularized by the philosopher Thomas Kuhn, in which the majority of scientists in a given field discard a traditional theory of nature in favor of a new one that passes the tests of experiment and observation; Darwin's theory of natural selection, Newton's gravity theory, and Einstein's general relativity all represented paradigm shifts.

Parsec—a unit of astronomical length equal to 3.262 light years.

Particle-wave duality—the fact that light in all parts of the electromagnetic spectrum (including radio waves, X-rays, etc., as well as visible light) sometimes acts like waves and sometimes acts like particles or photons; gravitation may be similar, manifesting as waves in spacetime or graviton particles.

Perihelion—the position in a planet's elliptical orbit when it is closest to the sun.

Perihelion advance—the movement, or changes, in the position of a planet's perihelion in successive revolutions of its orbit over time; the most dramatic perihelion advance is Mercury's, whose orbit traces a rosette pattern.

Perturbation theory—a mathematical method for finding an approximate solution to an equation that cannot be solved exactly, by expanding the solution in a series in which each successive term is smaller than the preceding one.

Phion—name given to the massive vector field in MOG; it is represented both by a boson particle, which carries the fifth force, and a field.

Photoelectric effect—the ejection of electrons from a metal by X-rays, which proved the existence of photons;

Einstein's explanation of this effect in 1905 won him the Nobel Prize in 1921; separate experiments proving and demonstrating the existence of photons were performed in 1922 by Thomas Millikan and Arthur Compton, who received the Nobel Prize for this work in 1923 and 1927, respectively.

Photon—the quantum particle that carries the energy of electromagnetic waves; the spin of the photon is 1 times Planck's constant h.

Pioneer 10 and 11 spacecraft—launched by NASA in the early 1970s to explore the outer solar system, these spacecraft show a small, anomalous acceleration as they leave the inner solar system.

Planck's constant (h)—a fundamental constant that plays a crucial role in quantum mechanics, determining the size of quantum packages of energy such as the photon; it is named after Max Planck, a founder of quantum mechanics

Principle of general covariance—Einstein's principle that the laws of physics remain the same whatever the frame of reference an observer uses to measure physical quantities.

Principle of least action—more accurately the principle of *stationary* action, this variational principle, when applied to a mechanical system or a field theory, can be used to derive the equations of motion of the system; the credit for discovering the principle is given to Pierre-Louis

Moreau Maupertius but it may have been discovered independently by Leonhard Euler or Gottfried Leibniz.

Proton—an elementary particle that carries positive electrical charge and is the nucleus of a hydrogen atom.

Ptolemaic model of the universe—the predominant theory of the universe until the Renaissance, in which the Earth was the heavy center of the universe and all other heavenly bodies, including the moon, sun, planets, and stars, orbited around it; named for Claudius Ptolemy.

Quantize—to apply the principles of quantum mechanics to the behavior of matter and energy (such as the electromagnetic or gravitational field energy); breaking down a field into its smallest units or packets of energy.

Quantum field theory—the modern relativistic version of quantum mechanics used to describe the physics of elementary particles; it can also be used in nonrelativistic fieldlike systems in condensed matter physics.

Quantum gravity—attempts to unify gravity with quantum mechanics.

Quantum mechanics—the theory of the interaction between quanta (radiation) and matter; the effects of quantum mechanics become observable at the submicroscopic distance scales of atomic and particle physics, but macroscopic quantum effects can be seen in the phenomenon of quantum entanglement.

Quantum spin—the intrinsic quantum angular momentum of an elementary particle; this is in contrast to the classical orbital angular momentum of a body rotating about a point in space.

Quark—the fundamental constituent of all particles that interact through the strong nuclear force; quarks are fractionally charged, and come in several varieties; because they are confined within particles such as protons and neutrons, they cannot be detected as free particles.

Quasars—"quasi-stellar objects," the farthest distant objects that can be detected with radio and optical telescopes; they are exceedingly bright, and are believed to be young, newly forming galaxies; it was the discovery of quasars in 1960 that quashed the steady-state theory of the universe in favor of the big bang.

Quintessence—a fifth element in the ancient Greek worldview, along with earth, water, fire and air, whose purpose was to move the crystalline spheres that supported the heavenly bodies orbiting the Earth; eventually this concept became known as the "ether," which provided the *something* that bodies needed to be in contact with in order to move; although Einstein's special theory of relativity dispensed with the ether, recent explanations of the acceleration of the universe call the varying negative pressure vacuum energy "quintessence."

Redshift—a useful phenomenon based on the Doppler principle that can indicate whether and how fast bodies in the universe are receding from an observer's position on Earth; as galaxies move rapidly away from us, the frequency of the wavelength of their light is shifted toward the red end of the electromagnetic spectrum; the amount of this shifting indicates the distance of the galaxy.

Riemann curvature tensor—a mathematical term that specifies the curvature of four-dimensional spacetime.

Riemannian geometry—a non-Euclidean geometry developed in the mid-nineteenth century by the German mathematician George Bernhard Riemann that describes curved surfaces on which parallel lines *can* converge, diverge, and even intersect, unlike Euclidean geometry; Einstein made Riemannian geometry the mathematical formalism of general relativity.

Satellite galaxy—a galaxy that orbits a host galaxy or even a cluster of galaxies.

Scalar field—a physical term that associates a value without direction to every point in space, such as temperature, density, and pressure; this is in contrast to a vector field, which has a direction in space; in Newtonian physics or in electrostatics, the potential energy is a scalar field and its gradient is the vector force field; in quantum field theory, a scalar field describes a boson particle with spin zero.

Scale invariance—distribution of objects or patterns such that the same shapes and distributions remain if one increases or decreases the size of the length scales or space in which the objects are observed; a common example of scale invariance is fractal patterns.

Schwarzschild solution—an exact spherically symmetric static solution of Einstein's field equations in general relativity, worked out by the astronomer Karl Schwarzschild in 1916, which predicted the existence of black holes.

Self-gravitating system—a group of objects or astrophysical bodies held together by mutual gravitation, such as a cluster of galaxies; this is in contrast to a "bound system" like our solar system, in which bodies are mainly attracted to and revolve around a central mass.

Singularity—a place where the solutions of differential equations break down; a spacetime singularity is a position in space where quantities used to determine the gravitational field become infinite; such quantities include the curvature of spacetime and the density of matter.

Spacetime—in relativity theory, a combination of the three dimensions of space with time into a four-dimensional geometry; first introduced into relativity by Hermann Minkowski in 1908.

Special theory of relativity—Einstein's initial theory of relativity, published in 1905, in which he explored the "special" case of transforming the laws of physics from one uniformly moving frame of reference to another; the equations

of special relativity revealed that the speed of light is a constant, that objects appear contracted in the direction of motion when moving at close to the speed of light, and that $E = mc^2$, or energy is equal to mass times the speed of light squared.

Spin—see quantum spin.

String theory—a theory based on the idea that the smallest units of matter are not point particles but vibrating strings; a popular research pursuit in physics for two decades, string theory has some attractive mathematical features, but has yet to make a testable prediction.

Strong force—see nuclear force.

Supernova—spectacular, brilliant death of a star by explosion and the release of heavy elements into space; supernovae type 1a are assumed to have the same intrinsic brightness and are therefore used as standard candles in estimating cosmic distances.

Supersymmetry—a theory developed in the 1970s which, proponents claim, describes the most fundamental spacetime symmetry of particle physics: For every boson particle there is a supersymmetric fermion partner, and for every fermion there exists a supersymmetric boson partner; to date, no supersymmetric particle partner has been detected.

Tully-Fisher law—a relation stating that the asymptotically flat rotational velocity of a star in a galaxy, raised to the fourth power, is proportional to the mass or luminosity of the galaxy.

Unified theory (or unified field theory)—a theory that unites the forces of nature; in Einstein's day those forces consisted of electromagnetism and gravity; today the weak and strong nuclear forces must also be taken into account, and perhaps someday MOG's fifth force or skew force will be included; no one has yet discovered a successful unified theory.

Vacuum—in quantum mechanics, the lowest energy state, which corresponds to the vacuum state of particle physics; the vacuum in modern quantum field theory is the state of perfect balance of creation and annihilation of particles and antiparticles.

Variable Speed of Light (VSL) cosmology—an alternative to inflation theory, in which the speed of light was much faster at the beginning of the universe than it is today; like inflation, this theory solves the horizon and flatness problems in the very early universe in the standard big bang model.

Vector field—a physical value that assigns a field with the position and direction of a vector in space; it describes the force field of gravity or the electric and magnetic force fields in James Clerk Maxwell's field equations.

Virial theorem—a means of estimating the average speed of galaxies within galaxy clusters from their estimated average kinetic and potential energies.

Vulcan—a hypothetical planet predicted by the nineteenth-century astronomer Urbain Jean Joseph Le Verrier to be the closest orbiting planet to the sun; the presence of Vulcan would explain the anomalous precession of the perihelion of Mercury's orbit; Einstein later explained the anomalous precession in general relativity by gravity alone.

Weak force—one of the four fundamental forces of nature, associated with radioactivity such as beta decay in subatomic physics; it is much weaker than the strong nuclear force but still much stronger than gravity.

X-ray clusters—galaxy clusters with large amounts of extremely hot gas within them that emit X-rays; in such clusters, this hot gas represents at least twice the mass of the luminous stars.